Investing in Dynamic Markets

Without venture capital, many of the companies whose technical innovations sparked the digital revolution would not exist. Venture investments funded these firms to develop their bright ideas into commercial products that created new business models and established whole new markets. In *Investing in Dynamic Markets*, Henry Kressel, a senior partner at multi-billion-dollar global investing company Warburg Pincus, takes you behind the scenes of the private equity process. He draws on his extensive experience to show how venture capital works, why venture capitalists fund certain companies and not others, and what factors influence the success or failure of their high-risk, high-reward investments. He also discusses venture capital's future, now that the commercialization of technology requires larger investments and global market access. Written in clear, non-technical language, the book features informative case studies of venture capital funding in a wide range of industries, including telecommunications, software and services, semiconductors, and the Internet.

HENRY KRESSEL is a senior partner of Warburg Pincus, LLC, where he has been responsible for investments in technology companies. He began his career at RCA Laboratories where he pioneered the first practical semiconductor lasers. He was the founding president of the IEEE Photonics Society and co-founded the IEEE/OSA *Journal of Lightwave Technology*. He is the recipient of many awards and honors, a fellow of the American Physical Society and the IEEE, was elected to the US National Academy of Engineering, and is the holder of thirty-one issued US patents for electronic and optoelectronic devices. He is the co-author of two previous books, *Semiconductor lasers and heterojunction LEDs* with J. K. Butler (1977) and *Competing for the future: How digital innovations are changing the world* with Thomas V. Lento (Cambridge University Press, 2007).

THOMAS V. LENTO is founder and President of Intercomm, Inc., a corporate communications consultancy specializing in technology companies. He has been a university professor, an ad agency executive, and Director of Communications for Sarnoff Corporation. In addition to collaborating with Henry Kressel on *Competing for the future: How digital innovations are changing the world*, he was editor of *Inventing the future: 60 years of innovation at Sarnoff*, and co-authored *By any means necessary: An entrepreneur's journey into space* with Gregory H. Olsen (2009).

Investing in Dynamic Markets

Venture Capital in the Digital Age

HENRY KRESSEL

THOMAS V. LENTO

CAMBRIDGE
UNIVERSITY PRESS

CAMBRIDGE UNIVERSITY PRESS
Cambridge, New York, Melbourne, Madrid, Cape Town, Singapore,
São Paulo, Delhi, Dubai, Tokyo

Cambridge University Press
The Edinburgh Building, Cambridge CB2 8RU, UK

Published in the United States of America by
Cambridge University Press, New York

www.cambridge.org
Information on this title: www.cambridge.org/9780521111485

© Henry Kressel and Thomas V. Lento 2010

First published 2010

Printed in the United Kingdom at the University Press, Cambridge

A catalogue record for this publication is available from the British Library

Library of Congress Cataloguing in Publication data
Kressel, Henry.
Investing in dynamic markets : venture capital in the digital age / Henry
Kressel, Thomas V. Lento.
   p.   cm.
ISBN 978-0-521-11148-5 (hardback)
1. Venture capital.   2. High technology industries.   3. Investments–
Computer network resources.   I. Lento, Thomas V.   II. Title.
HG4751.K73 2010
332′.04154–dc22   2010009138

ISBN 978-0-521-11148-5 Hardback

For Bertha
with love

Contents

Figures

Tables

Acknowledgments

My partners at Warburg Pincus and I owe an enormous debt of gratitude to Lionel I. Pincus and John L. Vogelstein for their remarkable leadership from the 1970s to 2002 in building the firm into a leading global investment institution. Co-Presidents Joseph P. Landy and Charles R. Kaye have led the firm since 2002, continuing its record of success.

Since the 1980s Warburg Pincus has grown both in the industrial specialties of the partners and in its locations, with offices in London, Frankfurt, Hong Kong, New York, San Francisco, Beijing, Shanghai, Tokyo, and Mumbai. It has invested over $29 billion in hundreds of companies located in thirty countries.

One cornerstone of the firm's success has always been its mastery in arriving at a true estimate of what constitutes an acceptable investment risk in the face of prevailing economic conditions, new industrial and technological developments, and the availability of outstanding entrepreneurial talent.

Another hallmark of Warburg Pincus throughout its nearly four decades of existence has been the teamwork that exists among the partners of the firm, and between the firm and the entrepreneurs in its portfolio companies. This collaborative approach, backed by an insistence on integrity of the highest order, inspires trust and enables the development of great new companies.

It was this pervasive spirit of collaboration that made my work so rewarding. Despite the international character of its operations, the firm operates as a single entity with a common purpose when it comes to promoting the success of its portfolio companies. The unique talents of partners globally located have enabled successful investments in many parts of the world.

Over many years, I have had the pleasure of collaborating with Joseph P. Landy, Dr. William H. Janeway, Charles R. Kaye, Jeffrey A. Harris, A. Bilge Ogut, Cary J. Davis, Steven G. Schneider, James Neary, Patrick T. Hackett, Kewsong Lee, Dr. Stanley Raatz, Barry Taylor, Stewart Gross, Mark M. Colodny, Dr. Frank Brochin, Chansoo Joung, Chang Q. Sun, Julian Cheng, Dalip Pathak, Dr. Joseph Shull, Jeremy Young, Dr. Harold Brown, Sidney Lapidus, Reuben S. Leibowitz, Robert Hillas, Beau Vrolyk, Christopher Brody, Andrew Gaspar, Henry Schacht, Julie Johnson Staples, Scott A. Arenare, Timothy J. Curt, Alex Berzofsky, and Patrick Severson.

Without outstanding entrepreneurs and managers, all of our efforts would go for nothing and this book would not have been written. Our success depends equally upon our ability to form strong partnerships with entrepreneurs capable of building interesting companies into great businesses. For the period of time covered in this book, I had the pleasure of working with Dr. Robert Pepper, Dr. Edgar Sack, Dr. Greg Olsen, Dr. Vladimir Ban, Edward Grzedzinski, Eli Harari, Jeffrey Ganek, Mark Foster, Mike Lach, Jack Hidary, Joseph Trino, Jeffrey Braun, Charles J. McMinn, Robert Knowling Jr., Will Wright, Richard Joyce, Atiq Raza, Behrooz Abdi, Mark Greenquist, Scott Melland, John Schneiter, Raviv Zoller, and Sachi Gerlitz.

Relationships with investment bankers are an important part of the investing and value creation process. I wish to thank Michael S. Wishart, now at Goldman Sachs & Co., for his valuable advice.

In writing this book, I have been fortunate in continuing my collaboration with Thomas V. Lento, who helped shape its content and without whom it would not have been completed. Valuable comments were provided by Dr. William H. Janeway, Dr. Peter Scovell, Dr. Stan Raatz, Jeffrey A. Harris, Bart Stuck, and Scott Meyer. Paula Parrish, the Commissioning Editor at Cambridge University Press, provided valuable guidance for refining its presentation. I am also grateful to Professor Ken Pickar of the California Institute of Technology for hosting my lectures which formed the basis of the case studies in the book.

Finally, I am very grateful to Tina Nuss for her extraordinary work in preparing and proofreading the manuscript.

Introduction

In early 2009 bestselling author and *New York Times* columnist Thomas Friedman wrote a column decrying the US government's decision to give failing American automakers billions of dollars in aid. He was writing in the wake of the worldwide fiscal crisis of 2008, the biggest since the Great Depression.

"When it comes to helping companies," Friedman says, "precious public money should focus on start-ups, not bailouts ... Let's make sure all the losers clamoring for help don't drown out the potential winners who could lift us out of this."[1] He goes on to suggest that top venture capital firms (VCs) pick the winners and allot the Federal funds.

It is clear that the importance of innovation in spurring economic growth and creating new wealth and prosperity is no longer in question. During the American presidential campaign of 2008 there were constant calls for the development of new, revolutionary energy technology as a way of reigniting the economy. But there was plenty of support for this position at the highest levels of government even before that. Here is an excerpt from the 2005 US President's "Economic report."

> Innovation is a primary engine of economic growth. Many commonplace features of modern life, such as personal computers, the Internet, e-mail, and e-commerce, have developed and diffused throughout the economy within a short span of years. Our Nation's growing prosperity depends on fostering an environment in which innovation will flourish.

---

[1] T. L. Friedman, "Start up the risk-takers," *The New York Times*, "Week in review" (February 22, 2009), 10.

The innovative process involves the invention, commercialization, and diffusion of new ideas. At each of these stages, people are spurred to action by the prospect of reaping rewards from their investment. In a free market, innovators vie to lower the cost of goods and services, to improve their quality and usefulness, and – most importantly – to develop new goods and services that promise benefits to customers. An innovation will succeed if it passes the market test by profitably delivering greater value to customers. Successful innovations blossom, attracting capital and diffusing rapidly through the market, while unsuccessful innovations can wither just as quickly. In this way, markets allow capital to flow to its highest-valued uses.[2]

However, such statements of faith in Adam Smith's invisible hand of the market romanticize the messy process through which technological innovations generate increased industrial and commercial productivity and fuel economic growth. Dr. William H. Janeway, a Warburg Pincus partner and respected economic analyst, puts it this way:

> The relationship between risk and reward remains as critical as it is problematic ... The history of venture capital is littered with examples of technological "solutions" in search of commercially identifiable problems; the history of venture capital is also littered with sets of "me-too" start-ups seeking to follow where others have already proven a market to exist. The point is that the selection of interesting investment opportunities requires the matching of evolving technological capabilities with evolving market needs.[3]

[2] *Economic report of the President* (Washington, DC: United States Government Printing Office, February 2005), p. 135.
[3] W. H. Janeway, "Doing capitalism: Notes on the practice of venture capitalism," *Journal of Economic Issues*, XX (2) (1985), 431–441.

In other words, it takes expertly managed risk capital to commercialize innovations. Experienced investors and entrepreneurs must come together to create new enterprises in a business arrangement that works to their mutual advantage. In a free enterprise system, technological innovations can be transformed into engines of economic growth only if those who take those risks have the prospect of reaping commensurate rewards.

In my view and that of many others, including Mr. Friedman, professionally managed venture capital remains the best model for that arrangement. Because new ventures are far more likely to fail than to succeed, actively engaged investors must play a key role to increase the chances of success.

And failures are an integral part of the venture capital investing process. In Janeway's words, "an economic system driven by technological innovation must have a robust tolerance for failure, an ability to absorb the waste of failed ventures."[4]

This book is intended to show how venture capital helps entrepreneurs profit from innovation through the commercialization of new technology. To accomplish that task it is divided into three broad topics:

- Chapters 1–3: how investment opportunities emerge, how venture capital works, and how industry dynamics influence new business opportunities;
- Chapters 4–7: case studies of thirteen venture-backed companies funded by Warburg Pincus in four different market areas, with summaries of the lessons each teaches about building new companies around innovative technologies;
- Chapter 8: analysis of key factors impacting the historical profitability of venture capital funds, particularly public markets, and likely future directions for investment.

Hopefully this will provide readers with a framework for understanding and appreciating the work of professional investors

---

[4] Ibid.

in dealing with the many factors that can impact the success of their ventures and hence the profitability of the funds they manage.

WHY VENTURE CAPITAL?

The general reader may wonder why anyone would write a book on a financial industry as specialized as venture capital investing.

My answer is that the need for venture capital is constant in modern economies, as it is one of the crucial engines of growth. The development of technological innovations to satisfy insatiable human needs and desires is not going to stop any time soon. As long as it continues, venture capital will be required to take the innovations to market.

That does not mean that the deployment of risk capital and its financial returns will remain constant. It will vary with prevailing economic conditions. Business cycles are a fact of life, affecting venture capital along with our other financial institutions. But, over the long term, economic growth will continue to be driven by commercialization of innovations and the creation of new markets.

Writing this book has also given me the opportunity to share my experience in venture capital investing. Since 1984 it has been my privilege to be a partner at Warburg Pincus, one of the largest and oldest investment firms. Warburg Pincus has a broad range of equity investing strategies to address many different industries, geographies, and stages of company development, from startup to mature business. The company traces its origins to the 1960s and raised its first fund in 1971. It is now in its tenth fund.

Venture capital funding has increased dramatically in the intervening years. This was a time when hundreds of new VCs were created in the hopes of profiting from the high returns delivered by this investment model.

Venture capital is a relative latecomer to the financing of innovation. Throughout the history of capitalism, when new technologies held out the promise of huge business opportunities, entrepreneurs and investors have always managed to find each other. In the early days, however, this was typically done through a financing process where entrepreneurs or freelance "company promoters" brought pools of investors together to back specific business opportunities.[5]

The well-known example of Thomas Edison provides a good illustration of how this was done in the US in the nineteenth century. His inventions were commercialized through *ad hoc* arrangements with investors. Today we would call this "angel" financing, but some of Edison's angels were bankers, including J. P. Morgan.

This form of financing continues to play a crucial role in underwriting early-stage "seed" companies, with billions invested in startups every year. But as technologies become more complex and more costly to develop, angel investing is generally no longer sufficient to take an innovation to commercial launch.

In the second half of the twentieth century the emergence of VCs, staffed with professional investors with backgrounds in industry, academia, and finance, changed the way technology-based businesses get created. With billions of dollars to invest annually, they funded thousands of new businesses.

The venture capital model is to create limited partnerships managed by general partners with the responsibility and authority to invest funds within broad guidelines. The institutions that provide these funds – the limited partners – are attracted by the promise of above-average financial returns, which they believe to be consistent with the risk of the businesses that venture capitalists finance. They count on the VCs and entrepreneurs to manage the risks.

MANAGING CHANGE

Though there were broader motivations for writing this book, including an ongoing need to better understand and communicate what drives the success of new technology ventures, my immediate impetus was a practical one. I needed case histories for discussion in

[5] N. R. Lamoreaux and K. L. Sokoloff, with a foreword by W. H. Janeway, *Financing innovation in the United States: 1870 to the present* (Cambridge, MA: The MIT Press, 2007).

a class on entrepreneurship that I present every year at the California Institute of Technology at the invitation of Professor Ken Pickar.

I started with a review of over thirty investments that my partners and I have made in technology businesses over a period of 24 years. I ended up selecting thirteen companies as representative of different kinds of investments, each of which presented interesting lessons about the process and outcome of venture development. The case studies of these companies are at the heart of this book.

The businesses that I selected include semiconductor and electronic equipment companies, software firms, and service organizations. They range from startups with nothing but a handful of hopeful entrepreneurs to companies that were spun out from larger businesses to become standalone entities. The investing period covered is the 1980s and 1990s, and these companies were participants in the age of digital electronics, the biggest growth technology of that era.

In reviewing these investments, I hope the reader will get a good insight into how professional investors sort out investment opportunities during periods when new markets are opening up and there are apparently limitless opportunities to be exploited by startup companies.

Some of these companies went public, with their stock traded on the leading exchanges, while others remained privately held and ended up being sold to larger entities. Some were US-focused, while others were international. I also include some examples of disappointing investments.

Looking back over many investments with the benefit of hindsight proved to be a fascinating experience. I found it particularly interesting to compare the expectations of my partners and me at the time of the investments with what actually happened afterwards. These patterns became apparent.

- Businesses rarely develop as planned. Sometimes an initial error in charting a business strategy becomes evident, but it need not be fatal. The fate of ventures is not written in the stars and timely action can result in a successful outcome.

- Chance and the ability to exploit unexpected opportunities both play a key role in the eventual fate of companies – if they are in the right markets and have strong leadership.
- Building business value requires close team effort between the investors and the management, usually over many years. Our successes resulted from tight partnerships with extraordinary entrepreneurs.

Beyond these observations a number of other patterns emerged that tended to be repeated from company to company.

The patterns I am talking about deal with "environment" and market issues that can change drastically for good or ill during the life of an investment. These issues must be understood as soon as possible, because they form the basis of changing a business strategy while there is still time.

In my experience, the constant in successful companies is their skill in managing change and innovation. This is a process where the investors and the management team must cooperate closely, because more often than not changes in company strategy require changes in expenditures. The ability to accurately assess the environment and quickly adapt to changes tests the ability of the CEO and his or her staff and is a key determinant of success.

Investors are very fond of focusing on "management quality" as a crucial determining factor in the success or failure of a business. Well, of course it is. But this is like saying that good weather improves the experience of a golf outing. It's obvious, but it doesn't cover all the variables.

While it is absolutely true that startups need outstanding management, more importantly they need the *right* management at the right stage of a company's development. It is important to install managers who can adapt to the challenges of the business as it goes through transitions. A startup in its early phases needs to have a very different management profile from that of a more mature company. A major task for entrepreneurs and investors is to understand the changing requirements and make sure that the right people are on board. That starts with the CEO.

Fast, flexible, and creative thinking is especially important in a CEO. Of all the skills and leadership qualities required of a CEO, a crucial one – and the hardest to find – is the judgment to react rapidly and effectively to new situations. Winston Churchill described the ideal military leader in terms that could just as easily apply to the head of a company in a highly competitive industry:

> Nearly all the battles which are regarded as the masterpieces of the military art … have been battles of manoeuvres in which very often the enemy has found himself defeated by some novel expedient or device, some queer, swift, unexpected thrust or stratagem. There is required for the composition of a great commander not only massive common sense and reasoning power, not only imagination, but also … an original and sinister touch which leaves the enemy puzzled as well as beaten …[6]

Here you have a crisp summary of a key skill that a CEO needs to lead a business to success.

ASSESSING RISK

Revolution, innovation, trend, fad, bubble: when we talk about technical progress and socio-economic change, even the language we use carries an undercurrent of uncontrollable risk and uncertain success.

If I can single out one factor that has led to business failure more than any other, it is overinvestment in emerging markets deemed to have unlimited growth possibilities. While venture capital funds have been responsible for many successful ventures, much capital has been spent funding ventures that eventually failed for any one of hundreds of reasons. Excessive optimism is one of them – perhaps the most common.

Entrepreneurs may have very diverse backgrounds but they share one key characteristic: they are optimists convinced that their business will succeed in the face of intense competition. Venture

---

[6] Quoted in A. C. Brown, *Bodyguards of lies* (New York: Harpers and Row, 1975), p. 5.

capitalists, the entrepreneur's partners, must share their vision, but they have to temper enthusiasm with realism, and weigh the risks of failure against the chances of success.

At least that is the ideal, but it is periodically discarded. For example, in the 1990s many investors lost their objectivity and poured seemingly endless streams of money into Internet startups. Investments in Internet-related companies hit an astonishing peak of about $79 billion in 2000 with 4,512 companies funded.

Everyone was acting like the party would go on forever. Just one year later, in 2001, the bottom dropped out. Overall investments dropped to $38 billion as stock market valuations crashed. By 2002 investments were down to only $9 billion, recovering to $28 billion in 2008. As a result, many startups failed after 2000 because they could not raise money.

The "dot-com" crash taught us once again that even the most far-reaching of innovations can be the basis for bad investments, particularly in early stages when applications and hence markets are unclearly defined. The result: a bubble, pithily defined by Eric Janszen as "a market aberration manufactured by government, finance, and industry, a shared speculative hallucination and then a crash, followed by a depression."[7]

It is easy to dismiss all investments made in startups in periods of rapid technological innovations as too risky. Of course this is not true, provided investors base their funding decisions on sober criteria. The reality is that successful investments were made during the dot-com period, just as they were in other periods of rapid technological change, by a selective process of evaluating opportunities and selecting good management for the companies being funded. That is the right way to create valuable businesses.

One way that investors tilt the odds in their favor is to invest in companies based on technologies and markets that they already

---

[7] E. Janszen, "The next bubble: Priming the markets for tomorrow's big crash," *Harper's Magazine* (February 2008), 39.

understand or have the background to learn quickly. Before joining Warburg Pincus I managed research and product development in electronics and optical communications in one of the leading R&D organizations in the world, the David Sarnoff Research Center of RCA Corporation. The Sarnoff labs pioneered many of the important industrial technologies, including color television, semiconductor lasers, and CMOS chips. When I became a venture capitalist it was only natural for me to focus on areas related to those fields, especially computing and communications.

Over the years I have been involved in funding companies during periods of extraordinary market transformation. As a bonus, I have participated in an industry that has done more in the past 50 years to improve lives and create social and economic progress than practically any other field of endeavor. That is the source of a good deal of personal satisfaction. But it was only possible because of close teamwork with my partners over many years. Teamwork is a core success factor at Warburg Pincus. It promotes the participation of many people in an investment as it evolves, leading to successes that no individual working alone could hope to achieve.

The investments discussed in detail in this book involved teamwork with my partners that extended over many years. I cannot overemphasize the importance of close teamwork in managing investment portfolios. My close collaboration with Joseph P. Landy (co-president of the firm since 2002) and Dr. William H. Janeway (formerly vice-chairman of the firm and now a senior advisor) began in the 1980s. In the 1990s, A. Bilge Ogut, Cary Davis, and Dr. Frank Brochin became closely involved in some investments. Over time other partners also made valuable contributions to investments discussed, including Jeffrey A. Harris, Charles R. Kaye (co-president of the firm since 2002), James Neary, Patrick T. Hackett, Steven G. Schneider, Mark Colodny, and Alex Berzofsky.

No investor group bats a thousand. Since failures often teach you more than successes, a number of disappointing investments are discussed. In each case we will explore the reasons for the original investment and for its ultimate outcome. The purpose of these discussions is to help the reader understand how we made our investment choices and the consequent developments that decided their eventual fate. Lessons learned will be highlighted, as they have general relevance for many other investments.

As we explore the world of venture investing, remember that we are dealing with a highly competitive process where a few winning companies emerge ahead of a host of other startups, all focused on similar opportunities in new markets. In this environment the biggest investment error of all is herd investing – that is, hoping to build a winning business by following the leaders with me-too strategies.

There is much to be learned from natural selection in the animal world. Some years ago I went hiking in the hills of Galilee in northern Israel. Suddenly, a goat appeared out of the brush ahead of me, with no herd dog or goatherd in sight. As I stood and watched, a herd followed. Behind them came the goatherd.

I engaged him in conversation: "How come the goats don't need close supervision? Don't they just stray without a dog?"

He smiled. "No risk of that happening. You see, there is always a lead goat, the smartest in the herd, who finds the way to the best food patches. The others just follow. All I have to do is come along behind. I have not lost a goat ever."

"If the first goats get the best food, what happens to the last ones in the herd?" I asked.

"Well," he answered, "they just take what they can get or go hungry."

And so it is in venture investing. The smart goats get the best returns. The others just make do with less or nothing. Losing money is their fate.

1 Hot markets, investment waves, bubbles, charlatans

> You want to spend $20 billion of taxpayer money creating jobs? Fine. Call up the top 20 venture capital firms in America ... and make them this offer: The U.S. Treasury will give you each up to $1 billion to fund the best venture capital ideas that have come your way. If they go bust, we all lose. If any of them turns out to be the next Microsoft or Intel, taxpayers will give you 20 percent of the investors' upside and keep 80 percent for themselves ...
>
> As we invest taxpayer money, let's do it with an eye to starting a new generation of biotech, info-tech, nanotech and clean-tech companies, with real innovators, real 21st-century jobs and potentially real profits for taxpayers.[1]

Venture capital firms (VCs) in the US have funded thousands of new entrepreneurial companies since the 1970s. By commercializing technological innovations, these startups have helped pioneer new industries, promoted job creation, and contributed to economic growth around the world. Their benefit to society is undeniable, as is that of the VCs that underwrote their growth.[2]

Most observers expect that venture capital will continue to play its crucial role in creating new engines of growth and prosperity. That is the conviction behind Thomas Friedman's call for the US government to stop propping up failing industries and use the existing US venture capital structure to build new ones instead. While this proposal is highly unlikely to find political support, it does reflect the recognition that VCs have earned over the years.

There are solid financial reasons for this awareness. Over the longer term, venture capital investments have had an impact on the US (and world) economy out of all proportion to their size.

---

[1] T. L. Friedman, "Start up the risk-takers," *The New York Times*, "Week in review" (February 22, 2009), 10.

[2] To avoid redundancy and confusion, we will use the term "venture capital" to refer to this type of funding. The abbreviation "VC" will be applied to the firms that manage the funding.

One study concludes that, in 2006, venture capital-backed companies were responsible for 17.6% of the US gross domestic product (GDP) and 9.1% of employees in all private companies. We are talking about 10.4 million jobs and $2.3 trillion in annual sales. Of particular interest is the impact of venture capital-backed companies in the electronics sector: about 78% of the revenues in computers and peripherals came from these companies, as well as 40% of all software revenues.[3]

Venture capital funding enabled the launch of major new markets. Giant companies such as Intel, Oracle, Cisco Systems, Apple Computer, eBay, and Google, all initially venture-backed startups, continue to play leading roles in those markets. Many lesser-known startups, still independent in 2009, also grew into industry leaders in their sectors. NeuStar (an investment discussed in Chapter 4) is a good example, and others are listed in Table A.1 in the Appendix.

Less understood are the contributions of venture capital-backed companies that seeded new markets and were then acquired because of their unique products or technologies. These companies made unusual business contributions, even though they no longer exist as separate entities. The list in Table A.2 in the Appendix includes a few examples, but does not do justice to the thousands of other companies that left a significant legacy before they disappeared as independent businesses. Their contributions are lost from view. By the time a company attains public status by issuing stock or being acquired, few people care to remember how it was started and what other firms it absorbed on the road to success.

While many ventures succeeded, many others fell by the wayside. Funding early-stage technology ventures is very risky, as we will have ample opportunity to discuss, and for this reason venture capital investors expect high rates of return on the funds committed.

---

[3] IHS Global Insight, *Venture impact: The economic importance of venture capital-backed companies in the US economy*, fourth edition, updated through 2006 (Arlington, VA: National Venture Capital Association, 2007).

Therefore it should not be surprising that the flow of new money into venture capital is critically dependant on its profitability, which has varied over time as economic conditions and financial markets have changed. As we will discuss at more length toward the end of this book, the market upheavals of 2000–2001 and later proved that venture capital investing is not immune to general business cycles or stock market cycles (which are NOT the same thing).

In the meantime we will explore how VCs choose the companies they invest in, the mechanisms for investing, and what factors affect the outcomes of their investments. Our findings will provide the context for the major part of the book: case studies of thirteen actual investments in companies in four different industries.

In this chapter we will present an overview of how venture capital investors respond to promising new opportunities in so-called "hot markets." These are exemplified by four important investment waves: small computers, telecommunications, the Internet, and alternative energy technology. (We will have a lot more to say about telecommunications and the Internet later in the book as we discuss specific investments.)

Once investment opportunities surface, the key question is, how can investors tell the jewels from the clunkers? In order to answer that question you first have to discuss how a new market opportunity impacts investment decisions.

By its very nature, venture capital investing is a matter of good timing. But it is also a process prone to excesses when enthusiasm for a new market outruns reality. Therefore, we start by looking at how investment waves emerge and how they can degenerate into bubbles, with unhappy consequences for all participants.

WAVES AND BUBBLES

Venture capitalists are in a race against the clock. They are deluged with hundreds of business plans from entrepreneurs passionate about their ideas. But they have a limited amount of time in which to consider these proposals and make new investments.

The reason is that venture capital funds have a finite life.

In contrast to private or corporate investors, the modern venture capital model is built around funds with a life limited to about 10 years. Their success is judged by the rate of return they make on the invested money. At the end of a fund's life, the assets need to be liquid and in a condition to be distributed to the limited partners (i.e., investors in the funds).

Therefore, most investments will be made during the first 5 or 6 years of a fund, leaving the rest of the fund's life as a "harvest" period. As a result, the average holding period for an individual investment is about 5 years. So it is essential that new investment opportunities be rapidly but carefully evaluated and closed as appropriate. Putting money to profitable use is the name of the game.

This also means that there is a great deal of competition among VCs in finding and closing on the best deals.

Given the competitive nature of the investing process, it is not surprising that some VC firms may fall victim to a certain amount of "group think" in picking investments. If certain high-growth opportunities suddenly emerge as "hot markets" because of new technologies or other reasons, investors are likely to race each other to be on the leading edge of investments.

The best recent example of herd investing was in the early Internet era. In the mid-1990s, as the Internet's potential became widely discussed and appreciated, startups emerged in ever-increasing numbers. At first these were financed by the best-informed VC firms, which had access to the best entrepreneurs. Soon other, less-qualified firms, afraid to lose out on a massive opportunity, felt compelled to fund Internet companies as well.

As a result, too many startups, led by immature management teams, piled into the same hot market. This is how a major investment wave, triggered by revolutionary innovations, becomes an irrational phenomenon leading to a bubble. The end result is an unusually high rate of startup failures, because few of the new companies are able to generate meaningful revenues.

Major investment waves like the Internet stampede are stimulated by a combination of revolutionary technologies and political or economic changes that people believe will create enormous new markets. We call these *waves*, not *cycles*, because each wave is unique, created by a different technology or set of events.[4]

Investment cycles, on the other hand, display a pattern of recurrence, with money gravitating to and flowing out of certain industry sectors as their fortunes ebb and flow. The steel industry is a good example. It periodically attracts new investment when business cycles create an increased demand for its products.

For the prudent investor the problem is knowing whether an investment wave reflects a sustainable growth market opportunity. Needless to say, prudence also requires understanding when too much money starts chasing a limited business opportunity and the wave turns into a bubble.

That's the problem with investment waves. Many culminate in a bubble. Bubbles grow out of mass hysteria, engendered by new markets that a large public believes to have unlimited potential for profit. In their excitement, investors cast aside traditional valuation methods and plunge money into any enterprise that is in the new game.

However, I don't want to minimize one important benefit of these periods of massive investment. They can drive the development of revolutionary technologies. While obviously excessive, investments in the dot-com era hastened the emergence of the Internet with all of its societal benefits. Investors simply overestimated its near-term economic potential.

Inflationary investing

Bubbles are not new. Most people associate them exclusively with the booms and busts of the industrial age, but their history is far longer.

---

[4] For an excellent historical review, see C. Perez, *Technological revolutions and financial capital: The dynamics of bubbles and golden ages* (Cheltenham, UK: Edward Elgar, 2002).

Three earlier bubbles were memorably described in a classic book, *Extraordinary Popular Delusions and the Madness of Crowds*, written by Charles Mackay in 1841. Mackay's account of the Dutch tulip mania of the seventeenth century describes a market in which people not only bought and sold tulip bulbs, but even speculated in tulip futures. The bubble burst in 1637, costing many investors everything they owned.

Some critics have accused Mackay of exaggerating the tulip craze, though they agree that his histories of the South Sea Company bubble of 1711–1720 and the Mississippi Company bubble of 1719–1720 are on firmer ground. Exaggerated or not, however, the fact is there was a tulip bubble. Silly as it seems now, the tulip frenzy demonstrates that bubble psychology is always the same. As Mackay put it, "men, it has been well said, think in herds; it will be seen that they go mad in herds, while they only recover their senses slowly, and one by one!"

What starts a bubble? Economists have proposed various theories, none of which offers a complete explanation. However, during my career as a venture capitalist, I have observed that bubble markets all exhibit several common characteristics, which become most evident in public company valuations.

First, bubbles occur during periods when abundant risk capital is waiting in the wings. Second, they require that major technology and market developments are under way. Third, these must catch the interest of a large investing public. And then, most important of all, public enthusiasm has to boost the valuations of public companies in related industries to extraordinary and unsustainable levels. These valuations become divorced from such historical metrics as profitability or revenues. Instead, they are driven primarily by the eagerness of new investors to own a piece of a business that they believe is guaranteed to be worth more than what they pay for it.

When all of these conditions exist, you have a bubble in full expansion. Unfortunately for investors, while bubbles expand slowly, they fail rapidly, as the mood of investors switches from optimism to pessimism. What was a rush to buy becomes a rush to sell, with new buyers nowhere to be found.

An investment wave officially becomes a bubble only after it is over and a great deal of money has been lost. While the run-up in investments is under way, optimists build elaborate financial models purporting to show that even the most outrageous valuation is justified by future business performance. Meanwhile, no one is paying much attention to the pessimists, who see a bubble forming in any sustained investment wave.

Or, as R. J. Samuelson put it, "When things go well, everyone wants to get on the bandwagon. Skeptics are regarded as fools. It's hard for the government – or anyone else – to say, 'Whoa, cowboys, this won't last.'"[5]

How do you know that you are in a bubble? The flip reply is that you don't until it bursts. The real answer is that in a true bubble anomalies abound, all of which are good indicators of coming grief. The first sign is a general belief that the market opportunities are boundless. The second is the total abandonment of historical metrics for company valuation, which are traditionally linked to profits or revenues.

Winners and losers

The last big technology bubble, created by the promise of the Internet and deregulated telecommunications, started to inflate in the late 1990s. The stock market performance of companies in these sectors tells the whole story.

On March 10, 2000 the technology-heavy NASDAQ stock index in the US peaked at 5132.52 in intra-day trading, double its value of just a year before (and five times its value barely 4 years earlier). A number of Internet companies had gone public by then

---

[5] R. J. Samuelson, "Good times breed bad times," *Newsweek* (October 27, 2008), 55.

and were enjoying huge valuations. It was a quick slide downhill from there. By March 15 the index was down to 4580, and the rout was on.

Between March 2000 and October 2002 the collapse of the public market, in what became known as the end of the "dot-com bubble," stripped all technology companies (and therefore their investors) of some $5 trillion in market value. That's a lot of money to lose in a very short time, and it set the stage for a recession. But the dot-com fiasco is hardly unique in its effect on the larger economy.

Earlier bubbles, such as the railway mania in Britain in the 1840s and the craze over automobiles and radio in the 1920s, also battered their respective national economies. And although the bubble of 2008–2009 formed around inflated real estate values and excess bank credit, it too followed all the classic patterns, and dragged down economies around the world in a way not seen since the US stock market crash of 1929.

Of course, not everyone takes a beating when a bubble pops. Early participants who manage to sell their holdings before the end of the bubble can make lots of money. Given how much money is lost when a bubble collapses, though, such lucky (or wise) investors are obviously in the minority. When the music stops, as it always does, values plummet and unhappiness prevails.

Any investment professional with a long enough career can tell stories of missing opportunities by not catching the wave early enough, or of losing money by waiting too long to exit. Some boast of their extraordinary genius in avoiding bubble industries and just watching others lose money.

Bernard Baruch, the financier and economic advisor to several US presidents, was famous for having sold all his stock just before the crash of 1929. He had read Mackay's book and saw all the signs of a bubble on the verge of popping. In our own era Warren Buffett, one of the most successful investors in modern history, avoided the dot-com disaster by simply sitting out the technology investing waves.

Technology makes waves

There is a critical link between technology evolution and investment opportunities. The reason is that most revolutionary technologies create new markets, giving rise to big investment waves as opportunities surface for new companies.

Existing companies rarely play a leading role in this process. It is usually newcomers that develop the revolutionary products which pioneer new markets. The incumbents who should have seen the opportunity sit on the sidelines.

Why do they miss the opportunities? Because big industrial leaders frequently lag in the commercialization of revolutionary technologies, particularly ones likely to obsolete their current products. When they do enter the new market, they do so late, or participate half-heartedly. Perhaps the best example that comes to mind is the replacement of film-based cameras by digital photography: Kodak, the leader in film technology, is not the digital camera market leader, although it is a participant.[6]

When an incumbent delays entering a new market related to its products, it leaves itself open to being outmaneuvered by focused and nimble new competitors. Eventually, however, when the big companies muster their substantial resources to enter the market, it sets the stage for intense competition – often through the acquisition of prominent VC-funded startups.

There is a common theme to big investment waves. They are all launched by visionary entrepreneurs. These are soon joined by investment bankers, industry analysts, and journalists who jump on their bandwagon. But the lure of new technologies and new companies with apparently unlimited potential can lead to irrational investments, with too many companies started simply to exploit a business opportunity at hand.

---

[6] For an excellent discussion of this topic see C. M. Christensen, *The innovator's dilemma: The revolutionary book that will change the way you do business* (New York: Collins Business, 2002).

Investors come to believe that the market opportunity is so vast that there is room for any number of successful new ventures. History shows that only a few big winners eventually emerge from any investment wave, but somehow the hopeful investor believes that his or her company will be among the survivors.

And let's not forget the external impediments to clear-sighted analysis that appear in every hot market. First is the emergence of charlatans touting phony science and fictional financial models to convince the gullible that they hold the key to progress and prosperity. The technological base of many modern investment waves gives these pitchmen an advantage. While the public could easily understand the tulip and railroad industries of earlier waves, sophisticated new technologies like software, electronic systems, and renewable energy are beyond the grasp of most investors.

This gives real and fake entrepreneurs alike the opportunity to solicit capital from audiences ill equipped to assess the technical or business merits of their claims. Early "angel investors" are particularly vulnerable. People know from popular lore that those who invested in pioneering technology enterprises right at the start often made a lot of money. Stories of huge fortunes made by early investors lure them into backing what they hope will be the next blockbuster company.

Investors who seek "sanity checks" find few independent experts they can trust. They flock to conferences hastily set up by self-styled experts on the new technology. These usually end up as promotional events. The hype targets professional investors first, but soon neophytes are drawn into the investor pool.

All that is needed to push eager believers to the irrational investment level are one or two examples of highly visible financial successes. Then it's open season on people looking for the big payoffs.

To gain a fuller appreciation of the impact of big investment waves, we will now review four waves that triggered many billions in venture capital and other investments.

These are all drawn from personal experience. I entered the venture capital industry in the midst of the big investment wave triggered by the rapid growth of personal and small business computers. I've survived the full gamut of the small computer, Internet, and telecommunications investment waves. Small computers did not trigger a bubble. The other two did. The fourth investment wave that we will mention, alternative energy sources, is ongoing.

1980s – WORKSTATIONS AND PERSONAL COMPUTERS

Few technologies in human history have had the economic impact of the personal computer (PC). If there was ever a product that could produce an investment frenzy, it was the PC. Introduced in the 1970s, in little over a decade it went from an exotic hobbyist preoccupation (how many people remember the Altair, let alone ever used one?) to a ubiquitous business tool and personal necessity.

IBM legitimized the technology when it introduced its first PC in 1981. By the middle of the decade PCs had shrunk in size, their cost had dropped, software applications had proliferated, and they were widely used in business. Figure 1.1 shows how the cost of the components used to build computers and other devices incorporating microprocessors declined dramatically starting in 1980.

While PCs were useful for everyday tasks like word processing and spreadsheets, they simply did not have the power to do big number-crunching tasks such as corporate payrolls, database management, or scientific calculations. Those applications were still the territory of "big iron" – the mainframe computers managed by IT departments. Mainframes were expensive to lease and operate. They required high-priced talent to keep them running, and even higher-priced experts to program them to do something useful.

Understandably, only very large companies had the resources to buy and support mainframes. Small and medium-sized businesses that needed computing capability wound up paying the

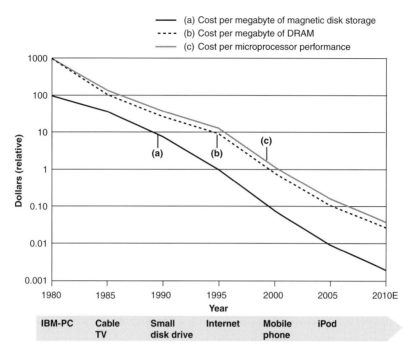

FIGURE I.I. Historical relative cost of three elements of computing systems: magnetic disk storage, DRAM, and microprocessor performance. *Source:* Mashey, TechViser.

big companies for the use of their machines under time-sharing arrangements.

What was needed was a more powerful small computer for general corporate use.

The emergence of the mid-sized computer opened a major new market. Digital Equipment Corporation, a highly successful company that emerged in the 1970s, scored a huge commercial success with minicomputers, a new class of medium-sized machines that could complement or even replace the big mainframes in many industrial applications.

By breaking big iron's stranglehold on computationally intensive applications, minicomputers spurred a rapid expansion in the market for computing machines. Now even small and medium-sized businesses could afford their own computers.

Continued progress in the field led to a fast-growing market for an even more affordable class of computers called workstations. As the logical next step in the evolution of the small computer, workstations once occupied an important place in the consciousness of the computer-savvy. Names such as Sun Microsystems, Silicon Graphics, and Apollo Computer loomed large as major players, along with IBM and Hewlett-Packard. There were debates about the merits of RISC processors and SCSI interfaces that only the initiated could follow. Every year a newer, more powerful class of machines appeared, offering greater capabilities for graphics, engineering, or scientific computing.

This growing market attracted a legion of entrepreneurs. The financial requirements for developing workstation products were modest, a few million dollars at most, so investors could get a piece of the action at relatively low cost. The biggest investment risk was whether a new entrant could develop a market for its workstations. Large enterprises were naturally wary of buying computing equipment from startups, because they worried about the potential for being stuck with orphan machines if the vendor failed.

When I joined Warburg Pincus in the early 1980s, workstation startups were the rage. In fact, so many new business plans for these companies came into my office that I set up a file named JAW (Just Another Workstation).

We passed on all of these business plans, for three good reasons.

- None offered a differentiated strategy in an industry that promised to have very few winning entrants.
- As Figure 1.1 shows, commercial components were steadily growing in performance, and their cost continued to drop. PCs would soon rival the workstations in capability while selling for much less.
- Established computer vendors were focused on the market, which meant that startups would face formidable competition from entrenched suppliers with well-established sales channels.

As we expected, the industry rapidly consolidated in the 1990s. A decade later the major providers of such business computers had been reduced to three: HP, IBM, and Sun Microsystems. Most of the others either went out of business or were acquired. There is no reliable record of the financial returns to the investors, although some VCs that had invested early in the period probably made good returns in some of these investments. Of the VC-funded startups only Sun Microsystems, a 1982 startup, survived as an industry leader for the next 20 years.

The PC business and the industries behind it – microprocessors and memory chips, disk drives, displays, even keyboards – represent one of the greatest and most successful investment waves of the twentieth century, or of any other period for that matter.

The workstation investment wave, on the other hand, has faded into history, and few of today's investors know anything about it. That's their loss. Understanding how market conditions and advances in technology ultimately turned workstations into a footnote in the computing industry can prepare an investor to be more critical about potential new opportunities in markets where similar factors are in operation.

1990S – THE INTERNET

The emergence of the Internet started one of the biggest investment waves of all time, and VCs contributed huge amounts of capital. As noted earlier, it did not arrive alone. The telecommunications wave, which we discuss below, formed – even co-evolved symbiotically – along with it.

While they did inflate into bubbles, both waves also left behind lasting benefits in the form of valuable technology and infrastructure. But that is a small consolation to the many folks who lost enormous sums when the two bubbles collapsed.

In fact, it can be argued that the benefits could not have been generated *without* the speculative excess and the losses generated thereby. By this logic a great deal of financially painful trial and error in business models – and in venture capital funding – had to occur for the winners to emerge.

The ascendancy of the Internet and telecommunications from perceived hot markets to waves to bubbles followed the classic pattern. When the Internet became accessible to the general public in the middle 1990s, people quickly grew enthusiastic about the birth of a new medium, and realized that its commercial impact could be huge. What was unclear was how the Internet could act as the vehicle for successful businesses.

Entrepreneurs presented plans to Warburg Pincus and others to use Websites to conduct commerce, distribute information, set up social networks (long before MySpace and Facebook) – the list was endless, and each potential new service attracted startups ready to commercialize them. Furthermore, all of them needed enabling software technologies to manage the communications and content.

All of the business plans assumed endless market growth free of difficult competition. They predicted huge increases in traffic and Internet users as more homes got connected, followed by more services to attract even more users. The Internet was touted as being as important as the revolution brought on by printing in the 1400s.

But the great majority of entrepreneurs lacked a coherent and workable business thesis for generating revenues and profits in the face of the prevailing public perception of the value of the Internet. Entrepreneurs' plans for monetizing the Web were up against serious market resistance from people accustomed to everything on the Internet being free. Though they could see the point of paying for a tangible product purchase or for software and services to enable the construction of Websites, few people were willing to pay for information delivered over the Internet.

Hence the common reliance on advertising-supported business models for startups built around delivering information and other content. Advertisers supported free content on radio and TV. Why not on Websites? Entrepreneurs vied with each other in their faith that banner ads would support any number of new startups offering free content.

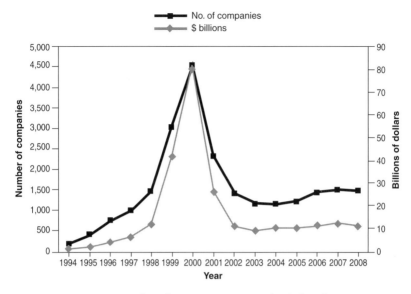

FIGURE I.2. Number of Internet companies funded with US venture capital and the total dollar amounts funded each year between 1994 and 2008. *Source:* Based on data from Thomson Reuters.

Investors were on board. Between 1994 and 2008, 22,868 Internet companies were venture capital funded to the tune of $244 billion (Figure 1.2). In the peak investment year of 2000, 4,540 companies received $79.7 billion, or nearly 80 percent of all of the invested venture capital. The appeal of the sector has since cooled markedly, but even in 2008, when 1,478 companies were funded, there were still lots of investors eager to bet on the 'Net.

Today the Internet has fulfilled the most outrageous prediction of its early champions. It has indeed transformed industries and services. However, the number of successful Internet companies created in the 1990s is but a small fraction of those funded. Giant successes such as Google, Yahoo!, Amazon.com, and eBay are exceptions rather than representative models. Far too many startups were funded for the market opportunities available, and many of them failed.

While many investors lost a great deal of money on these ventures, others actually came out ahead. Some startups became public companies before being acquired, even though they were not profitable. Others were acquired for their technology at valuations that made their investors and managers rich. The most prominent example is Skype, which had practically no revenues yet was acquired by eBay for $2.6 billion in 2005.

Some Internet startups became viable mid-sized companies. In Chapter 6 we will discuss in detail two Warburg Pincus investments as examples. The first, GlobalSpec, remained private as an industry leader in its sector. The second, EarthWeb, became public but changed from an unsuccessful advertising-supported model to a successful subscription model, and changed its name to Dice.

1990S – TELECOMMUNICATIONS

Telecommunications was already in the middle of a major restructuring when the Internet emerged. The two sectors became linked because the Internet greatly accelerated the transformation of the telecommunications industry. Although originally separate investment waves, by 2000 they had merged into a two-chambered bubble, with the fate of one chamber inextricably linked to that of the other.

Because Internet usage was producing an enormous increase in data traffic in the US and around the world, the rallying cry of those promoting telecommunications as a hot investment area was that the infrastructure would need to be greatly expanded to meet the burgeoning demand. They based their analysis on wildly optimistic forecasts of future needs.

In the 1990s industry analysts went so far as to claim that data traffic on the backbone networks was doubling every month. They were wrong. The real numbers were much lower, as later data showed.

But at the time such false estimates fueled enormous interest in investments in communications equipment suppliers. Public valuations of companies such as Lucent and Nortel reached stratospheric numbers. The valuation of Lucent hit $238 billion in 1999, while that of Nortel reached $206 billion in 2000. Cisco's valuation reached $485 billion in 2000.

Many venture-backed startups entered the market with specialized components (such as lasers), digital systems, and software for managing digital network traffic. Quite a few had their initial public offerings and enjoyed huge valuations as well.

New communications services providers entered the market with large appetites for new equipment but no ability to pay for them. To keep revenues flowing, some equipment vendors loaned these service providers – their customers – billions of dollars in the form of financing to acquire their equipment and software. It was a classic conflict of interest on both sides, and investors did not understand what was happening.

Investor interest was already waning by 2000, as industry revenues and profitability failed to meet analysts' outlandish expectations. This is best illustrated by the stock market index for telecommunications, shown in Figure 1.3.[7]

As the chart shows, the index peaked in mid-2000 and dramatically dropped thereafter. Investors were rushing for the exits, and a stampede toward the bottom was on. The bursting of the telecommunications bubble alone wiped out over $2 trillion of asset value worldwide. Most of the new service providers went out of business when their customer bases failed to grow fast enough to cover debt and operations.

Again, not everyone lost out. Some fortunate startups were acquired by large companies, resulting in large profits to VCs and other investors. Between 1996 and 2000 Cisco acquired sixty-three companies for a combined price of $33.4 billion. Lucent acquired thirty-eight companies for a total price of $3.45 billion.[8]

---

[7] Courtesy of Dr. A. Bergh (2005), Optoelectronics Industry Development Association (OIDA); data from BigCharts.com.

[8] W. R. Koss, *Six years that shook the world* (New York: BookSurge Publishing, 2006).

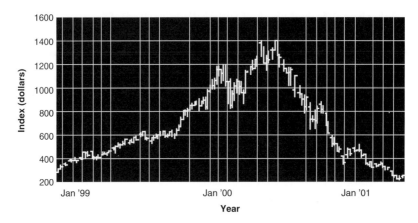

FIGURE 1.3. The bubble in the US Telecom Networking Index, October 1998 to October 2001. *Source:* BigCharts.com, quoted by Bergh (2005). Copyright © 2008 Marketwatch, Inc. All rights reserved (ref. 7).

At Warburg Pincus we committed funds to telecommunications companies early in the investment wave. I have selected five of these investments for detailed discussion in Chapter 4, where I outline the factors that contributed to their outcomes.

2000S – ALTERNATIVE ENERGY SOURCES

Computers of all sizes, telecommunications, and the Internet are still very much with us as forces in the world economy. As big investment waves, however, they no longer capture headlines. We have another giant in the making.

"Green technology" is the current hot topic among investors. It shows every sign of building into an investment wave that is likely to last for years.

While economists and policymakers have been talking for years about the need to find alternatives to fossil fuels, the run-up in oil prices since 2000 and concerns about climate change have made the search for "clean" energy a global concern. Media events have helped on this score. For example, Al Gore's Oscar-winning *An Inconvenient Truth* helped to spur the US government-sponsored push to encourage the development of renewable sources of energy through subsidies.

In short, we may have reached a tipping point in public attitudes over the past few years. Until recently people and governments have been, if not negative, at least passive about alternative sources of energy.

- The "oil shock" of the 1970s sounded some alarm bells, but as prices dropped and supplies became plentiful once more, the old complacency returned.
- Americans have been reluctant to expand nuclear energy facilities, considering the technology too unsafe as well as too costly.
- Other renewable sources of energy, such as those derived from wind-driven turbines or solar energy, were deemed too expensive for all but remote locations devoid of other energy supplies.

As long as oil is cheap, none of these renewable energy sources can attract the capital necessary for its exploitation without a massive government subsidy.

But a year or two of $4 per gallon gasoline and $100 per barrel oil got everyone's attention. Today, with oil still expensive and future pricing trends pointing upward, the public is adamant about preparing for a future in which America is less dependent on foreign oil, and is a net exporter of clean energy technology. Both political parties recognized the new attitude by making energy and the environment an issue in the 2008 presidential campaign.

Given this sea change in public attitudes, it is not surprising that venture capital investments in alternative energy companies have been on the rise. The relationship between the price of oil and VC investing is shown in Figure 1.4.

From a low of $235 million in 2003, VC investments in alternative energy companies rose to more than $4 billion in 2008. Over the same period oil prices went from just over $30 a barrel to a peak well above $100.

While slackening demand due to the recession and market collapse of 2008–2009 has driven oil back down to the

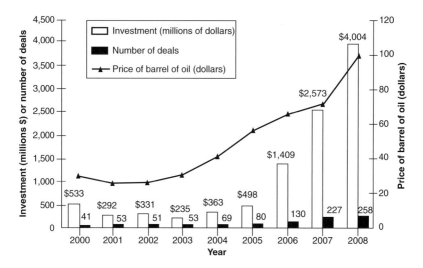

FIGURE 1.4. US venture capital investments in alternative energy companies since 2000. Also shown is the average price of a barrel of oil. *Source:* Based on data from Thomson Reuters; oil data from Bloomberg Finance LP.

$50–$70 range as of this writing, the US administration is continuing to push for alternative energy projects. Indifference to alternative energy sources is a thing of the past. Alternative energy sources have become a multibillion dollar market which will have its ups and downs driven by public policy, energy use, and the price of fossil fuels.

Yet even if investors are ready to climb on board an investment wave focused on alternative energy, there are still some obstacles in the way of financial success. Two frequently mentioned technologies – biofuels and solar power – testify to the issues entrepreneurs and investors face in such investments.

Biofuels

Humans have been burning natural materials for fuel for eons, from brush and wood to dried cow dung. The technology to turn corn or sugarcane into liquid fuels has been around for many years, but the cost differential between these fuels and oil pumped from the ground has been too high to make them economically interesting.

It took US government subsidies to attract private capital investments in ethanol production plants. Now that those plants are finally in place, their products must compete with oil-based fuels or present their investors with operating losses.

There are significant risks, starting with fluctuating oil prices. In addition, investments dependent on government largess must often contend with the risk of economic distortion, not to mention policy inconsistencies and erratic public perceptions.

In the case of ethanol, for example, press reports highlight the fact that the expansion of ethanol production is not only destroying forests around the world, it is creating food shortages too, as corn and sugar cane is diverted from food consumption to fuel production. (Such concerns are likely to be temporary.)

It has also been claimed that the chain of production of biofuels using current technology increases the emission into the atmosphere of the detrimental gases that are blamed for global warming.[9] How such arguments will impact continued government subsidies is an open question.[10]

Before leaving this subject, however, let's not underestimate human ingenuity. Other solutions may exist. It may well be possible to develop more promising biofuel technology using plants other than corn or other currently known sources. This is a new field which is sure to get venture investment looking for better solutions.

Solar technology

Criticisms of biofuel policy make solar energy sources look more interesting as a field for development.[11] And indeed, solar energy technology has gained enormous support from the venture capital community.

[9] G. F. Will, "The biofuel follies," *Newsweek* (February 11, 2008), 64.
[10] K. Allison and S. Kirchgaessner, "Investors suffer as US ethanol boom dries up," *Financial Times* (October 22, 2008), 1.
[11] J. Carey, "The biofuel bubble. The world is awash in startups working to replace fossil fuels and corn ethanol with better biofuels. Most will fail and big oil may steal the show," *Businessweek* (April 27, 2009), 039–042.

But once again the investment interest is unlikely to continue without government subsidies. Reporting on the growing solar industry, *The New York Times* noted in an editorial, "Investors are unlikely to pump much more money into clean power unless they are sure the [tax] credits will be available next year."[12]

Solar is often lumped together with wind as promising, non-polluting sources of renewable energy. Both have a lot of public appeal – except when the wind-driven generators are near proponents' homes! In comparison to solar, however, wind power generation technology has one big disadvantage from a VC's point of view. Up to now it has been largely the province of big corporations such as Siemens and General Electric. Some new companies have emerged internationally: Vestas in Denmark, for example, has become an industry leader in turbines.[13] Still, the field is dominated by major companies.

The appeal of solar energy generation is obvious. It has non-polluting and non-destructive; its energy source regenerates itself every day at sunrise; and it does not require the diversion of food crops into fuels.

As a technology, solar is pretty straightforward. Harnessing solar energy is done in two ways:

- the heating of fluids to sufficient temperatures to run a conventional gas turbine;
- the use of solar (photovoltaic) cells, which are semiconductor devices that convert light into an electrical current.

Scientists and engineers have been refining these approaches since the 1970s and have made enormous progress toward cost reductions.

Solar cells, the best-known devices for solar energy generation, have provided the primary source of electricity on space satellites for decades. They are successfully used in remote locations or aboard

---

[12] "Clean power or dirty coal," *The New York Times* (February 10, 2008), 11.
[13] A. Stone, "Blade runner – Ditley Engel is staking the future of Vestas, the world's largest turbine maker, on America. Watch for stormy politics," *Fortune* (April 27, 2009), 104–112.

boats to recharge batteries. Over time, solar cells have even found increasing use as complements to grid power, but not as substitutes for it. Big cost gaps (and inherent differences) exist between solar generation of electricity versus the use of fossil fuels.

The cost gap will not disappear anytime soon. But solar suffers from other inherent limitations as a mainstream source of energy. Production of solar power is limited to daytime, and output power can fluctuate depending on the weather and on seasonal variations in sunlight. It takes large numbers of solar panels to produce meaningful amounts of energy, which translates into a requirement for extensive open space not always available in the right places. That is why roofs and desert areas are such attractive sites for solar energy generation. Nor are solar cells useful in all climates. In spite of these constraints, if the cost of the panels can be reduced, there is an opportunity for creating a vibrant market for solar energy as a supplementary source of power useful during peak daytime utilization periods.

In manufacturing terms the production of the photovoltaic cells used for solar panels is related to the semiconductor industry, although the technology is very different. There is great hope that major improvements in efficiency at lower device costs are possible if the right talent is focused on solving the manufacturing problems. That is where the opportunity exists for venture companies.

Given that it is still uneconomical to produce large amounts of energy from solar panels, however, government support is essential to enable serious commercial use. Legislation and subsidies are needed to maintain a large market for solar cells, both by encouraging their use through tax rebates, and by forcing the power utilities to buy solar-sourced electricity at prices above commercial rates.

Germany, which in 2008 consumed half of the world's solar cell production, is a prominent example of such government support. Other places where solar energy sources are subsidized include Spain and the state of California. More countries are in various stages of providing subsidies.

It still took a potent combination of rising oil prices (Figure 1.4) and government subsidies to produce a wave of VC-funded startups formed to manufacture solar cells. Solar cell manufacturers use different technologies in the competitive game of making their devices cheaper than those of their rivals. The dominant technologies are still centered on the use of silicon. Innovations continue to lower the cost of the cells while increasing their efficiency.

The production of solar cells has become a global activity, with major suppliers in Germany, Japan, and the US funded by corporations as well as venture capital. The total global solar cell device market is now large – over $9 billion in 2008. In fact, since 2000 a number of new solar cell companies have gone public at enormous valuations, in anticipation of high growth rates and an expansion of the solar cell markets to countries ready to subsidize their use.

With over one hundred solar cell technology companies funded by venture capital since 2000, this has been an active investment wave. But as was true with biofuels, investments in solar energy face non-technological risks: fluctuations in the price of oil and government subsidies that are at the whim of policymakers. Investors in the second half of 2008 saw large drops in the public market valuations of solar energy companies when the price of oil retreated below $50 a barrel.[14]

As is typical in hot markets, dubious publicity is fueling the investment interest. Here is an e-mail I received on January 15, 2008 from an unknown source touting an obscure small-company stock. This is a good example, sloppy grammar and all, of how irresponsible promotion of obscure companies is used to spark irrational investments.

> ATLAN INTL HLDGS CP (ATLI.pk): The hottest company in
> "ON FIRE"Industry for 2008 Renewable Energy

> No one can escape the rising oil prices, will prices reaching
> record highs for oil in 2008, it's no surprise that renewable

---

[14] C. Krauss, "Alternative energy suddenly faces headwinds," *The New York Times* (October 21, 2008), B1.

energy companies are on FIRE as more and more people
stake their bets on cleaner sources of energy.

All solar companies in 2007 have returned over 500% and have
reached ridiculous valuations.

We have found an undiscovered gem, with long standing record
of pro active renewable energy research and development

ATLAN INTL holdings.Sym : ATLI

They have divested all subsidiaries not involved in Alternative
energy, increasing their liquidity and cash flow. Currently
@ $.25 and news expected in coming weeks we believe its
a good time to accumulate for a high reward breakout in trading
volume and price. Furthermore, Atlan has global exposure with
presence on the Frankfurt exchange as well

This is the time to be in Solar and in ATLI.

This, by the way, is a company that had just announced it was
divesting all subsidiaries not related to the alternative energy
industry. Its previous business? Nutritional supplements and live-in
Alzheimer's centers.

Let the investor beware.

GAUGING ENTREPRENEURIAL INTEGRITY

Investing in young companies that are entering new markets and
commercializing technological innovations is obviously a risky
proposition.

We are talking about diving into uncharted waters, and there
are many questions that must be considered before committing funds.
Will the technology work as planned? Will there be a market for it,
and when will that develop? Can the team that we are investing in
get its act together in the face of inevitable competition? How much
capital will be needed before the business becomes self-supporting?

The list goes on, as we shall see when we return to this topic in
Chapter 3. When trying to sort through such issues one has to rely on

information provided by the entrepreneurs and consultants, and on information found in the media and government research reports.

Understanding the risks is difficult enough if one is dealing with entrepreneurs of high integrity. Unfortunately, as all professional and angel investors eventually discover, this is less common than is generally believed. Hot markets breed charlatans all too eager to promote the equivalent of perpetual motion machines. This is not a new problem. In the fourteenth century John of Rupescissa, a Franciscan, complained that there were too few genuine natural philosophers because "most of those who pretended to pursue the sciences were magicians, sorcerers, swindlers and false coiners."[15]

You hear claims of magical new processes or new products that obsolete known solutions with better performance and lower cost. Then there are the self-proclaimed visionaries who paint a roadmap to the future and convince the public to follow it through repeated broadcasting via the public media. A healthy dose of skepticism is always wise.

Here is an example from my own experience.

At Warburg Pincus we were approached by a group of technologists (let's call them Team X) proposing to build a solar cell factory. It would use a new technology based on a chemical process that they claimed produced solar cells at half the cost of existing commercial processes.

In fact, the chemical compound they were using was not new, but Team X claimed that they had discovered a solution to some long-standing problems that had plagued earlier efforts to use it in cell production.

My search of the technical literature confirmed that respected laboratories had made many attempts to commercialize this material in the past. The results had always been promising, in that the best devices produced by this process had very high efficiency.

---

[15] F. Heer, *The medieval world – Europe 1100–1350* (London: George Weidenfeld & Nicolson Ltd, 1961), p. 239.

The problem was process variability: The results were so unpredictable that commercial production was impractical. Furthermore, the devices made with this process were extremely sensitive to ambient conditions. Their operating lifetime was heavily dependent on the environment in which they were used.

While this history was not encouraging, new solutions frequently surface that can transform the value of a previously impractical technology. So it pays to listen to new ideas.

Team X included PhDs formerly employed in a highly regarded laboratory, along with other staff with extensive semiconductor device experience. They had hired an investment banker to find investors for their new company.

The company had created a slick book with pictures of a pilot production plant and curves showing the performance of the solar cell. It was supported by the report of an independent consultant indicating that his findings supported the data shown. Also included was a lengthy description of the estimated future revenues of the new company and its profitability, based on its low costs.

Team X certainly presented well. Because of this we arranged to visit the facility.

We found that the claims were false hopes without experimental foundation. Only a few devices had ever been made, and they were produced in a small-scale laboratory. The pictures of the pilot plant were misleading as well. Nothing had ever been made there. In fact, the equipment they had in place could not possibly produce the devices they were proposing to make at the costs discussed.

Furthermore, we found that most of the production equipment that would be needed to manufacture such devices at the proposed scale simply did not exist. It would all have to be invented to produce the volume of product needed.

Some might debate whether the claims of Team X were knowingly falsified, or simply a highly optimistic picture of the future. But these are mere quibbles. The presentation from Team X was obviously designed to mislead investors anxious to enter the hot solar energy business.

I call that fraud.

GOAL: RETURN ON INVESTMENT

Venture capital investing is not a simple process of just funding a promising, high-growth company, sitting back and waiting for it to succeed, and then picking up a tidy profit when you finally sell your ownership stake.

If only it were that easy. The job of the venture capitalist is to identify and invest in promising companies and build a portfolio of investments whose potential financial returns justify the risk. Individual partners are usually responsible for several such investments as they progress through their lifecycle. Important decisions must be made frequently, if not daily. There is never a dull moment as each portfolio company goes through its periods of successes and problems.

But before we analyze the venture capital investment process, it is helpful to review the history of venture capital, and the principles under which VCs operate. This is the subject of our next chapter.

2 Financing high-risk businesses[1]

> In the morning sow your seed and in the evening do not be idle,
> for you cannot know which will succeed.
>
> (Ecclesiastes 9–6)

Large-scale professional venture capital is a relative newcomer in the world of finance. It emerged as a modern industry in the US in the 1970s, when a number of professionally managed limited partnerships were formed in response to the lack of other institutional funding for new companies. Their purpose was to finance businesses that balanced high risk against great potential for growth and profitability.

Throughout the history of capitalism, investors have been willing to accept the risks of investing in the commercialization of innovations in the expectation of unusual financial rewards. Whether the technology was steam engines, railroads, electrical power, or even canals, it has always been possible to raise capital for venture companies created to bring innovations to market, provided the potential was large enough.

Entrepreneurs of the past typically financed their new businesses with their own money, plus investments from family and friends, corporations, wealthy individuals (called "angel investors"), and customers. Banks were not a viable source of capital for startups in the US because Federal laws from the 1930s restricted them from investing in certain classes of assets.

Bootstrapping a business in this way, without institutional support, is a very difficult process. Microsoft and a few others have shown that it can be done. But technology companies generally require substantial up-front capital to develop a product, build

---

[1] Certain material in this chapter has been adapted and updated from H. Kressel with T. V. Lento, *Competing for the future: How digital innovations are changing the world* (Cambridge: Cambridge University Press, 2007), pp. 176–214.

facilities and gain market attention, and in most cases such private financing does not provide enough capital.

Whether entrepreneurs go for institutional backing right off the bat, or wait until they have proven their concept, they will eventually need funds to grow their companies. At that point it makes sense to turn to a specialized funding organization that manages a pool of money to finance promising businesses.

From the entrepreneur's point of view, the ideal funding organization would also provide the basis for a stable, long-term relationship between investors and entrepreneurs. This is particularly important to new companies in rapidly growing markets, where follow-on investments are usually needed to expand a business beyond the startup stage.

Professional venture capital filled that void, providing an organized approach to funding new enterprises. It reached maturity at exactly the point in history when the big corporate labs, which had been responsible for developing many of the basic electronic technologies and bringing them to market, were forced by global competition to focus their research and development (R&D) on near-term products. This opened an opportunity for startups to jump-start new markets.

Professional venture capital does more than give entrepreneurs a way to finance the growth of their companies. Those who invest through the venture capital funds also have the opportunity to benefit from high-growth industries. Large holders of capital looking for profitable investments, such as pension funds, soon recognized that they could reap attractive financial returns if they invested in businesses focused on commercializing new technologies and services. These institutions also realized that they needed specialists to manage this investment process, since it was likely to be as complex and fast moving as the underlying technologies.

Over the years, as we shall see in Chapter 8, the best venture capital funds delivered superior financial returns. But, there were big differences between the best funds and the rest of the industry in terms of performance.

HOW VENTURE CAPITAL BECAME A GROWTH INDUSTRY

To appreciate the impact of venture capital on the economy, consider how fast it became a primary source of funding for entrepreneurial companies.

The earliest notable venture capital company (VC) was American Research and Development Corporation, active from 1946 to 1973. Another prominent firm, Greylock Partners, was founded in 1965. It is still active in the industry, having raised $500 million in 2005 for its twelfth fund.

Warburg Pincus opened its first venture fund in 1971 with $41 million of committed capital. Over the years it has broadened its investment strategy. In 2008 it raised its tenth fund, for $15 billion.

Since the founding of pioneers like these, the industry has expanded at a tremendous rate. The number of venture capital management firms in the US grew from 89 in 1980 to 741 in 2007. Their professional staff also mushroomed, from 1,335 in 1980 to 8,892 in 2007.[2]

Overall investments also ballooned, with the peak years coming between 1999 and 2001. Figure 2.1 tracks investments made into VCs between 1979 and 2008 by institutions. The amount of money put to work by these funds is about equal to the amounts taken in.

The fact that the amount of venture capital available after 2000 has declined is the result of complex factors related to the profitability of the funds, a topic which is discussed in detail in Chapter 8.

While we will focus on venture capital funds, other sources of private risk capital should also be noted. They sometimes complement each other by financing businesses at various stages of development.

---

[2] *National Venture Capital Association (NVCA) yearbook 2008* (New York: Thomson Financial, 2008), p. 15.

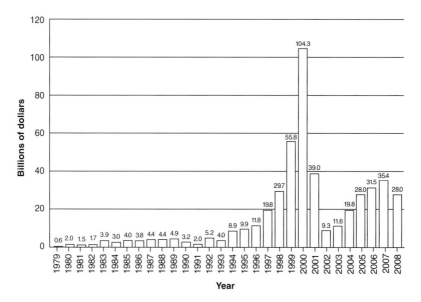

FIGURE 2.1. Capital commitments to US venture capital funds by financial institutions, 1979–2008. The amount of money invested each year in companies approximates the amount taken in each year by the funds. *Source:* Based on data from Thomson Reuters.

ANGEL INVESTORS

Although professionally managed venture capital funds have become a crucial financial resource for young companies, they have not replaced entrepreneurs' need for early-stage financing by wealthy individuals.

These "angel investors" are often the only source of startup money available for sums under a million dollars. Most VCs are now so big that it would be impractical for them to manage such small investments. Angels, on the other hand, specialize in seed stage funding.

They are frequently organized in informal investment groups that pool their resources for specific deals. Though the individual investments may be small, they add up to a surprisingly large total. A study by the University of New Hampshire's Center for Venture Research indicates that individuals invested an estimated

$24 billion in startups in 2008. This is nearly comparable to the total amount of venture capital investment for that year.[3]

Given how small the typical angel investment is, this translates to the funding of tens of thousands of new companies. No one knows how many last more than a year or two, but venture capital funds commonly finance the promising survivors.

Angel investors provide an invaluable service. As seed investors, they help get the highest-risk companies off the ground: startups which otherwise would not be funded at all. Once given a chance, some do emerge as successful companies. Microsoft actually started with such angel investor funding.

However, angel investors are fickle. Their appetite for risk is affected by the prospect of public market exits, the fear of being diluted in subsequent funding rounds, and their belief in "hot markets" where big, fast profits can be anticipated. Internet startups, which need only small sums to get started, have been and continue to be favorite angel investments.

INDUSTRIAL HOLDING COMPANIES

An industrial holding company owns a substantial percentage of the voting stock in one or (usually) more firms. It uses this ownership to exercise control over management and operations, which it can do by influencing or electing a company's board of directors.

There are many well-known holding companies. Berkshire Hathaway, the company headed by Warren Buffett, owns GEICO, Executive Jet, Dairy Queen, and many others. Loews Corporation is the parent company of Loews Hotels, Lorillard, CNA Financial, Diamond Offshore Drilling, and more.

While industrial holding companies and VCs both own equity in businesses, there are crucial differences between them.

---

[3] Whittemore School of Business & Economics, University of New Hampshire, Center for Venture Research: http://wsbe.unh.edu/cvr-news. Another good source of information on angel investment is the Ewing Marion Kauffman Foundation.

- Holding companies generally invest in established businesses, while "classic" VCs mostly focus on startups and young companies. Occasionally, holding companies co-invest with VCs in anticipation of a future acquisition.
- Holding companies can own assets for indefinite periods of time, while venture capital and private equity funds have a finite life – usually 10 years. That is the typical timeframe within which capital must be returned to the original investors.

PRIVATE EQUITY AND VENTURE CAPITAL FIRMS

The equity investment model of large firms like Warburg Pincus spans a number of strategies, so drawing hard distinctions is difficult, but an attempt at clarification will be helpful. Most firms denoted as "private equity" have different investment strategies from "venture capital" ones. Here are some differences.

On one end of the private capital investing spectrum are venture capital firms, which typically invest in companies that are in the early stages of development. Very young companies constitute risky investments, because neither their technology nor their business strategies have been market tested. To manage the risk in such early-stage investments, it is common to have several funding rounds which finance a company's growth and cover operating losses in stages, until positive cash flow from operations is achieved. After some years, such companies end up with a consortium of investors as successive rounds of funding attract new venture capital financing.

On the other end of the scale are "buy-out" firms. These are organizations that typically use a combination of equity and debt to acquire ownership of their portfolio companies. A typical buy-out transaction would be the purchase of a business with a history of steady, multimillion-dollar revenues and sustained profitability. Institutions such as banks help finance these acquisitions on the basis of the cash flow generated by the acquired company.

Buy-out and private equity

In recent years, responding to the negative publicity generated by the buy-out boom of the 1980s that culminated in the collapse of Drexel Burnham, a powerful investment bank, buy-out firms have successfully appropriated the term "private equity," as if the sole style and strategy for making illiquid equity investments were represented by the "leveraged buy-out" or LBO.

The classic LBO model is to repay the borrowed funds over time from the cash flow of the business, and then sell it. Alternatively, the exit might be through a public offering where the shares are sold over a period of time to the public. The strategy of the investors is to focus on improving the profitability of the business while expanding sales by significant changes in its operations.

Companies acquired with lots of debt have included big manufacturing businesses of all kinds, leading semiconductor companies (such as Freescale), retail store chains such as Neiman Marcus, and even large equipment companies such as Avaya. Obviously, the availability of debt at favorable rates is critical in making this model work.

From the point of view of this book, the critical factor is that dependence on innovation – and the risks necessarily associated with commercializing it – is a big factor in assessing the suitability of a business for acquisition through an LBO. The reason is that continued funding of innovation requires substantial reinvestment of earnings. Repaying large amounts of debt, as must be done in most LBOs, makes such reinvestments difficult. In a choice between legal obligations to debt holders and funding product development, legal obligations will always take precedence. Hence, excessive debt is undesirable.

In between LBOs and funding startups lies a spectrum of investment strategies. These range from late-stage growth investing in new companies that have passed the critical early tests of commercial success, to "distress" investing in turn-around situations.

Venture capital strategies

A VC's investment strategy depends on the amount of capital it manages and the interests of the partners. Some firms specialize in particular industry sectors, such as communications, software, biotechnology, or alternative energy. Another area of specialization is according to the maturity of the companies they fund. Some prefer to handle early-stage companies, while others fund enterprises that have already demonstrated a degree of financial and operational stability.

VCs are staffed by partners with extensive industrial and financial experience that mirrors their firm's investment strategy. They have the industry contacts to point them to good ideas, and the expertise to select appropriate candidate companies for their firm's investments. The general partners who manage the funds have broad discretion in investment decisions.

How much money is available varies widely from one firm to another. Some VCs manage as little as $20 million and limit their investments to seeding startups. Others, with over $1 billion under management, have broader strategies that include providing expansion capital to more mature companies.

The compensation structure is reasonably uniform across the industry. The operating "general" partners of the venture firms receive an annual management fee tied to the size of the funds they invest, as well as a share of the profits realized by the investments.

Because the investments are in privately held companies, they are illiquid. It normally requires several years of development before a private company's value can be realized through a sale or a public offering of its shares. That is why the typical life of a venture capital fund is at least 10 years, at which point the fund is liquidated and its assets are returned to the limited partners. There are provisions for a longer liquidation period. However, VCs generally do not receive additional fees when the life of the fund is extended.

While a fund may run for 10 years, it holds the companies in its portfolio for a much shorter period of time. This is because a VC raises money for a fund, not for specific companies: the investments a fund makes are chosen after the money is raised. As noted in the previous chapter, the median age of a venture capital investment at liquidation is generally between 5 and 6 years.

Firms that span various investment strategies

Before leaving this topic, I should note that the historically clear distinction between "classic" venture capital and "classic" LBO firms has, in some cases, become blurred. In this book, when we talk about venture capitalists we are describing professionals in the business of investing in technology-based growth companies who are actively involved in their development. "Financial engineering" is not the determining factor in the creation of the value of the businesses we discuss in this book.

Since we will be covering Warburg Pincus investments exclusively, a word about the firm is in order. Warburg Pincus uses different models to invest globally in a broad range of companies and industries. The firm's investments historically have included both the classic venture capital and LBO models depending on the industrial sector, geography, and characteristics of the individual opportunity. Given the size of the funds under management and the broad range of industry-specific skills among the firm's partners, the investments can range from startup companies to large enterprises.

The thirteen investments that we discuss in this book include such classic venture capital investments as WSI, discussed in Chapter 5, and examples of what we at Warburg Pincus call "growth equity" investments, such as Nova Corporation, discussed in Chapter 7. By "growth equity" I mean investments where Warburg Pincus is the primary capital source of the business and commits to a strategy of providing incremental funding as the successful growth of a company warrants the infusion of new capital.

Serving as the principal investors in such companies carries a big responsibility. This strategy requires a very close working relationship between the Warburg Pincus partners and the entrepreneurial team over a long period of years. *Partnership* is the key word to describe a relationship that leads to success for the benefit of all parties.

An important point to remember: financial information on the investment funds and firms that we discuss in this book (Chapter 8) refers only to those denoted as "venture capital" ones by the reports of the National Venture Capital Association. Data for firms that are significantly in the LBO category are not included.

HOW VENTURE CAPITALISTS MANAGE INVESTMENTS

Venture capitalists view their mission as the creation of valuable businesses through active participation in all phases of the development of their portfolio companies. That clearly calls for involvement far beyond simply putting money into a company. They have to engage in activities that help give each company its best chance for success.

What do VCs do in the course of this work? Much depends on the investment strategy that their fund pursues, but venture capital activities may begin with the idea of forming a business and extend all the way to its final sale. Here is a summary that reflects the range of potential involvement, from uncovering an opportunity to engineering an exit.

- Finding investment opportunities takes creativity. Therefore, an important element of venture capital work is to identify technologies and/or companies that show promise in creating markets with unusual growth potential. This calls for meeting hundreds of people and companies. Finding a needle in a haystack is probably the best analogy.
- Suppose that some promising ideas surface and the fund is interested in creating a very early-stage company around a new technology. The VCs then focus on finding management teams to build the business. This is a particularly important step for VC firms that like to invest in early-stage companies by collaborating with university faculties.

- VCs receive hundreds of proposals. Entrepreneurs often approach VCs with ideas that are promising but require a great deal of refining prior to investment. The task here is to work with entrepreneurs to formulate a business plan based on the perceived market opportunity.
- Once VCs decide to invest, they take up the difficult task of negotiating investment terms with the entrepreneurs and earlier investors (if any). This includes agreeing on a board of directors, structuring options for the entrepreneurial team, and settling governance issues. These are all fundamental to the successful operation of the business.
- The hardest part of venture capital work begins after the investment closes. Expectations and reality diverge quickly, and VCs must make a sustained effort to track the company's progress and to work with the company to fix problems, up to and including making changes in strategy or in management. In fact, growing companies face constant pressure to keep pace with the necessary management skills. Ensuring that the appropriate talent is recruited and trained is one of the most important aspects of venture capital activity.
- Among the most urgent problems that face young companies is funding. VCs actively work with management of the company to raise fresh capital as needed and to build corporate relationships that open improved routes to markets or easy access to new products. Part of this task is to identify opportunities for acquisitions.
- Since all venture capital investments call for an exit sooner or later, VCs work with management on positioning the company for an initial public offering (IPO) or sale. Here the challenge is to build corporate value. Factors such as superior management, product leadership, revenue growth, and profitability are key elements in value building.

What a list!

Clearly VCs cannot be passive investors. Because of the constant demands of managing a high-tech company in unpredictable and competitive markets, overseeing investments requires unwavering attention and involvement, plus the good judgment that can only be gained from experience.

In addition, VCs must have the legal power to make strategic and managerial changes in their portfolio companies in order to carry out their fiduciary responsibilities. The vehicle through which they exercise their prerogatives is the company's board of directors.

Given the variety of work that venture capitalists need to do, it is not surprising that there are wide variations in the skills of the partners in various firms. This talent is eventually translated into financial returns. Perhaps the biggest difference between firms, and one that greatly impacts the comparative profitability of funds which we discuss in Chapter 8, is the quality of the deal flow.

It is a system with strong feedback loops. VC firms that acquire a reputation for integrity and strong deal leadership – and successful investments – will attract the best entrepreneurs, and the best entrepreneurs give VCs the best chance of launching successful ventures.

WORKING THROUGH A BOARD OF DIRECTORS

The board of a startup usually includes representatives from VC firms with major stakes, the chief executive officer of the company, and experts selected for their industry knowledge. Outside members play an extremely important role, as they bring a broad range of pertinent experience onto the board.

Typically, such boards do not exceed six to eight people, who normally meet monthly for a detailed review of all activities of the business. This structure provides the oversight and control VC firms need to respond quickly to anything that might impact the company's chances of success.

With a new company, every day brings new challenges and opportunities for entrepreneurs and investors to address. Should the company launch a basic product to generate revenue, as investors might wish, or hold off until it has the comprehensive feature set envisioned by the entrepreneur? Can all of the technology be developed internally, as the innovator would prefer, or must the company acquire some of it from the outside to move the product to market faster?

There is never as much time to make decisions as one would like, or enough information on which to base them. Yet the board and management must make those decisions, or risk having random events control the future of the business. This is only possible in an atmosphere of trust, where entrepreneur and investor alike believe their needs and aspirations will be addressed in a fair and pragmatic way.

INCENTIVES FOR ENTREPRENEURS

Great entrepreneurs are independent minded. It is reasonable, then, to wonder what attracts outstanding entrepreneurs and management teams to start or join venture capital-backed companies when they know they will be under constant scrutiny from their investors and their board.

The first reason is that true entrepreneurs are not afraid of risk and seek opportunities to build great companies with the right financial partners.

The second incentive is the potential for large earnings. This derives from the fact that a venture capital-backed company offers its entrepreneurs and employees significant equity in the business. The distribution of ownership varies widely from one company to another, depending on each company's history and the total amount of equity invested. It is not unusual, however, for employees to own 10–25 percent of the economic value of a company at the time the investors sell their stake, either following an IPO or through the sale of the business.

If the company is successful, those employees can share many millions of dollars when the company is sold or has a public offering of its stock. Intel is a prominent case of how early employees can become millionaires after their company's IPO. This happy example has not been lost on entrepreneurs and technologists in Silicon Valley and elsewhere, many of whom have attempted to replicate it, with varying degrees of success.

Options, incentives, and taxes

Employees usually acquire ownership in startup companies through grants of stock options. While only 3 percent of all US companies offer stock options, they have been part of the employee incentive program in virtually every venture capital-backed company.[4]

Stock options give employees a chance to share in the capital appreciation of a business without incurring any direct cash cost to the company or its employees. In this way they help a company reward talented people while conserving cash for growth and development.

The practice of issuing stock options has been so successful that many high-technology companies continue it even after they go public. However, in the interest of total disclosure to investors of anything that might affect present or future corporate earnings, the US Department of the Treasury and the Internal Revenue Service have imposed regulations that force companies to place a value on these options.

Valuation of options is highly problematic. Methodologies for doing so, legally required under the new regulations, have unattractive financial reporting implications for both public and private companies. A company offering options to its employees has to record them as an expense. This affects the company's reported profits for the current year, thereby reducing its perceived value, even though there is no cash transfer to the employee. It is noteworthy that institutional investors in public companies have effectively required such companies to break out the non-cash charges generated by the granting of options so that investors can see through them to the underlying cash profitability of the business.

THE ENTREPRENEURIAL SOCIETY

Venture capital relies on entrepreneurs to create and build market-innovative companies. VCs were originally US based, because that is where both the technological opportunities and the liquid capital looking to fund their commercialization were concentrated. But VCs have

---

[4] B. McConnell, "Venture firms fret that a job creation bill could curb the use of stock options," *The Deal* (September 25, 2005), 20.

since proliferated around the world, especially during the bubble years of the late 1990s. In light of this development it is worth reflecting on the factors that influence popular interest in entrepreneurship, and to what extent it, too, is now a global phenomenon.[5]

We should first note that the US has no monopoly on entrepreneurs. People with the talent and drive to create new technology, new businesses, or new industries can be found everywhere.

Still, a large number of foreign entrepreneurs do set up shop in the US. In fact, recent immigrants make up a surprising percentage of the founders of US-based technology companies. One study shows that close to 30 percent of the Silicon Valley companies started between 1984 and 2000 were founded by immigrants.[6]

Immigrants also constitute a large fraction of the engineering staffs at some US technology companies. Some will eventually leave to start ventures of their own.

The fact that there are so many foreign-born and, in some cases, foreign-educated entrepreneurs and innovators proves that the entrepreneurial spirit operates across national borders. So-called "national character" is not the decisive factor in promoting or discouraging risk-taking. It is more accurate to speak of "national culture."

Why do these immigrants choose America as the place to pursue their dreams?

The answer is that it offers them a unique set of cultural and business conditions. A big factor favoring the US is the attitude toward failure. In Europe and other parts of the world, where examples of successful startups are scarce, failure in business carries very unpleasant social consequences. It can brand an individual for life. The fear of failure is a huge deterrent to entrepreneurship in these countries.

---

[5] D. G. Messerschmitt and B. Stuck, "Effective communications: The what, why and how of entrepreneurship," *IEEE Signal Processing Magazine*, 105 (July 2008), 1053–1056.

[6] R. Florida, *The flight of the creative class* (New York: HarperBusiness, 2004), p. 107; derived from A. Saxenian, *Silicon Valley's new immigrant entrepreneurs* (San Francisco: Public Policy Institute of California, 1999).

To give one example of this mindset, sociological studies suggest that when people in the UK with the skills to build businesses are presented with entrepreneurial opportunities, they are much less likely to take the leap than their US counterparts.[7] In fact, until quite recently, individuals who were directors of a company that declared bankruptcy in the UK were barred from ever again serving on the board of a business enterprise.

In the US, on the other hand, business failure does not carry the same social stigma, and it certainly does not end a career. Instead of being shunned by their peers, entrepreneurs with a failure in their past are often given credit for surviving a learning experience. The assumption is that this will help them be more successful in their next venture.

Entrepreneurship in the US has had another stimulus: the 1990s breakup of large companies such as AT&T (for the second time), ITT, W. R. Grace, GM/EDS, and Host Marriott, and the changes in or disappearance of other former industry leaders, such as General Electric, Eastman Kodak, RCA, and Xerox.

The turmoil of that era produced a generation of employees who no longer believed that they could gain job security simply by working for a Fortune 500 company. They had seen too many people, including themselves, lose their jobs for reasons unrelated to their skills or performance.

Once these talented employees had decided that lifetime jobs with a single company were a thing of the past, it was much easier to attract them to new, smaller companies that offered exciting work. In essence, they were willing to take larger risks for bigger rewards, because working at established companies was not all that safe anyway.

Europe and Japan have yet to experience a similarly massive restructuring of their own major companies, and that may be one

---

[7] See R. Harding, *Global entrepreneurship monitor* (London: London Business School, 2004), and J. Moules, "The start-ups that finish badly," *Financial Times* (November 16, 2005), 9.

reason why the urge to strike out in a new venture is less common in those areas of the world. However, given the entrenched nature of the local culture in most countries, it is unclear whether even the disillusionment that follows such a huge layoff would stimulate entrepreneurship the way it did here.

In fact, entrepreneurship is not universal even in the US, in spite of the favorable conditions we have been discussing. Instead, it is a regional phenomenon. For example, entrepreneurs enjoy great local prestige in California, Boston, and other areas where there have been a significant number of successful startups. They are looked on as role models for others.

In other parts of the country, where big companies have historically dominated local industry, the situation is quite different. People may talk about entrepreneurship, but they still see it as a less attractive option than joining an existing company for better job security.

Still, entrepreneurial companies are appearing in formerly unlikely places outside the US, and venture capital is following the opportunities. Warburg Pincus, headquartered in New York City, began developing a global strategy in the 1980s with offices and partners in Asia and Europe. It is now taken for granted that technology firms must have a global outlook, and investors are acting accordingly.

WHERE VENTURE CAPITAL IS INVESTED

Regardless of geography, VCs seek out businesses within their investing horizon that have the potential for unusual value creation. These tend to be in dynamic markets driven by rapid technology changes.

In 2008, $28.4 billion was invested in 3,192 companies. Not surprisingly, many venture investments continue to be in the high-technology areas of digital electronics, software, and communications. More recently, companies focused on alternative energy generation, power savings, and related technologies have received

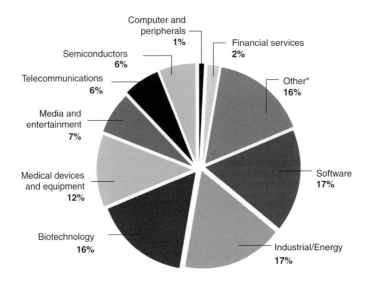

* Other includes: healthcare services; retailing and distribution; consumer products and services; business products and services; electronic and instrumentation; information technology services.

FIGURE 2.2. Venture capital investments in 2008 – by industry sector. *Source:* Based on data from Thomson Reuters.

increased funding since 2000 (see Figure 1.4, p. 32, and related text in Chapter 1).

Figure 2.2 shows the distribution by industry sector. Software received 17% of the money, while 16% went into biotechnology, and 6% into telecommunications and semiconductors, respectively.

Industrial and energy investments received 17% of the funds. Within this group, alternative energy technology investments grew roughly 56% to $4 billion from 2007 levels (see Figure 1.4).

HOW DO YOU VALUE A COMPANY FOR INVESTMENT?

I have frequently been asked how VCs decide on a valuation for a business they are considering for funding. Theoretical discussions do not give the true flavor of the process that investors undertake in evaluating new investments, so we will illustrate it with an example from real life.

I noted earlier that venture capital-funded startups usually go through several investment rounds to raise the money to fund growth. New funds are typically invited to join previous investors in order to distribute the investment risk.

In early 2008, Warburg Pincus was approached by a semiconductor startup that had been in business for 2 years and was just beginning to ship product. The company had three VC investors who had already invested $40 million in the company, and it was ready to solicit additional investors for a new round.

The first product was completed on budget and had a few customers. Revenues in 2007 were $5 million, but the company was unprofitable and would remain so until revenues grew to an approximate annual rate of $100 million, due to heavy development costs. If its new products were successful, management projected revenues of $800 million in 2012 and a net income of $80 million. The company was seeking new investors for an incremental $20 million which it believed would be sufficient to carry it to profitability.

The original VCs were asking for a valuation of $120 million with the new money included. Too high a price going in makes a profit at exit more difficult, if not impossible. Was that investment priced right?

The first point of analysis is assessing the probability that the company can reach its predicted financial performance. If a new investor becomes convinced that this *is* highly probable, then a future value calculation can be made. Predicting future valuations is the tricky part of the analysis.

If the investor assumes that chip company valuations in 2012 will be within the high historical numbers, then this company's public market valuation in 2012 – assuming "normal" market conditions – would be about twenty times its net income, or about $1.6 billion. On this assumption, an investor coming in at a valuation of $120 million would be getting about thirteen times his investment back *if his ownership percentage did not change because of additional money raised by the company in the interim.*

This is the point at which the investor has to assess various risks, any one of which can reduce such prospective returns.

- The company may need to raise more capital because of delay in reaching positive cash flow, thus diluting the investor's ownership.
- The company's revenues and profitability may be lower than projected.
- Economic and public market conditions may prove to be such that the anticipated company valuation is not reached. If the market for IPOs is not open for the company at the anticipated exit time, the valuation put on the business by a corporate acquirer, if one indeed exists, will reflect the absence of the IPO option.

On the other hand, the investor could get lucky and good things could happen to boost the company's value.

- Revenues and profitability could grow faster than anticipated.
- A corporate buyer might need the technology and be willing to pay a big premium for the company even before it goes public.
- Public markets for fast-growing technology companies may blossom again. We could return to a really frothy period when a multiple of ten times revenues – not net income – is the 2012 market standard. This was actually the case during some periods of the 1990s, which will be discussed in Chapter 8. The company would then be worth $8 billion, assuming that it reached its revenue target. Wow!

Deciding on such an investment is all about careful analysis of markets, weighing risk factors, and estimating a young company's prospects and management team. The investor's experience and industry knowledge are important in making such judgments. In the next chapter we review factors that enter into such decisions. In the end, after risks and opportunities are weighed, the decision to invest or pass is made.

The *decisive* issues are estimating (1) when the venture can become a self-sustaining, cash-positive business; and (2) how much equity capital will be needed to reach that point. Once the company has achieved this status, of course, one can afford to wait for the IPO market ... except that the longer the wait, the greater the risk of not maintaining a competitive position.

In any case, investors need to constantly emphasize "positive cash flow" versus "profitability" and the need to escape the tyranny of requiring outside (private or public) equity.

Just for the record, we passed on this chip company investment. It just did not meet our investment tests. We reasoned that the company's products and management were good, but it was competing in a market that was likely to stymie its growth long before it reached $800 million in revenues.

3 Venture investing: An uncertain science

The business cycle consists in essence in the ebb and flow of innovation, together with the repercussions resulting therefrom.[1]

Exiting an investment is the moment of truth for venture capitalists. That is when they learn the real value of the enterprise they have been funding. They measure its success by the multiple of the original investment they receive on exit.

The path from the start of an investment to the exit point can be long and tortuous, and forecasting the future value of a young company is a very uncertain science. A host of internal and external factors can conspire to hinder or promote a technology company's progress over the several years it takes to mature. But venture capitalists chance it anyway, trusting that their experience in selecting and managing investments and their active involvement in the company's development will improve its chances of success.

Understanding the environment in which a prospective portfolio company is expected to prosper is an essential part of the venture capitalist's job. This chapter discusses some of the major factors that affect technology companies and their markets. Although the information is general in nature, it serves to set the stage for the case studies we will consider in the next four chapters.

We begin with the fundamental driver of all technology businesses: the innovations they are bringing to market. We will examine obstacles to successful commercialization, talk about how companies shield innovations from competition, and gauge the risks that confront specific new technologies on their way to market. Then we will turn our attention to powerful market forces that can make or break a young company regardless of the nature of its innovation.

[1] A. H. Hansen, discussing J. A. Schumpeter in *Business cycles and national income* (New York: W.W. Norton & Company, 1951), p. 301.

ASSESSING TECHNOLOGY RISKS

Every investment decision is predicated on a set of assumptions about how much capital the business will need to become financially self-sufficient. Obviously, such assumptions involve a careful assessment of technology risks.

By technology risk assessment I mean estimating the degree of likelihood that the company can produce the desired products, on the predicted schedule, and at the anticipated cost. This helps establish how much capital the company will need to attain key milestones in creating marketable products.

It is fairly easy to make that assessment when the company is planning to introduce some well-understood adaptations to proven technologies, such as integrating digital camera circuitry into a mobile phone, to produce its proprietary new product. In this case most of the risk is related not to the technology, but to its market acceptability relative to competitive offerings. And if it is that easy to develop the product, there will be lots of competitive offerings!

However, when investors are faced with a truly radical innovation, it is even more important for them to assess the viability of the technology. It is also significantly harder, because there is no prior commercial application or market history on which to base a judgment about its commercial potential.

In this case, venture capitalists must evaluate not only the uncertainties of a new market, but the potential of a new technology and how long it will take to turn it into a marketable product. Here the biggest unknowable risk factor is timing.

A matter of timing

Uncertain timing translates into needing an unpredictable amount of capital to fund a new business. That is why some professional investors shy away from investments in companies using a radically new technology. They see them as long shots at best.

If this seems a little short sighted, it is because most people are not aware of how long it takes for revolutionary innovations to create big, profitable markets, or how short the reign of early pioneers at the top of a new market can be. Let's look at the history of four revolutionary technologies, each of which eventually created huge markets.

- transistors to integrated circuits (ICs): 20-plus years
  - 1947: germanium transistor invented;
  - middle 1950s: finally finds a meaningful market in transistor radios;
  - early 1960s: silicon transistors and ICs make germanium transistors obsolete;
  - 1970s: silicon IC chips come to dominate electronic systems.
- optical communications: about 20 years
  - late 1960s: modern semiconductor laser, the fundamental device in optical communications and consumer electronics such as CD and DVD players, is a laboratory curiosity. Low-loss optical fibers are demonstrated;
  - 1970s: lasers finally go into commercial production based on a variety of new structures developed over the intervening years. Improved optical fibers are developed along with low-cost production methods;
  - 1970s: fiber optic communications systems demonstrated;
  - 1980s: mass market develops for optical communications after enormous improvements in the quality of the fibers and related electronic systems and components.
- liquid crystal displays (LCDs): 20–25 years
  - late 1960s: flat panel LCDs, which now dominate the computer display and television receiver market, are first developed in primitive form;
  - 1970s: production techniques developed to manufacture reliable devices at low cost;
  - 1980s: commercial maturity reached, with very complex formats enabling portable computers and color displays;
  - early 2000s: bigger LCDs largely replace vacuum picture tubes in TV receivers and other cost-sensitive displays.
- light-emitting diodes (LEDs): 30-plus years
  - 1960s: red light-emitting diodes first demonstrated;
  - 1970s: they begin replacing tiny light bulbs used for instrument indicators;

- early 2000s: blue LEDs become available. It becomes possible to produce low-cost white color-emitting devices, and LEDs begin to replace conventional devices in more general use.

As noted, each of these revolutionary innovations took 20 or more years to move from invention to commercially successful products. Their long progress to high-volume manufacturing could not have been funded by venture capital, not just because it took so long to get them to market, but also because the cost of moving them from laboratory prototypes to finished products was so high.

It cost hundreds of millions, if not many billions, of dollars to commercialize each of these inventions, far more than most venture capital firms (VCs) would be willing to commit. This work was underwritten by big corporations such as AT&T, IBM, Corning, Sharp, Siemens, and RCA, and for the most part was carried out in their laboratory facilities.

Undeterred by long development timeframes and other startup risks, entrepreneurs and investors continue to be attracted by the commercial promise of revolutionary early-stage innovations. Success is elusive, but some great companies do owe their existence to such high-risk investments.

Unfortunately, more early pioneers fall by the wayside than succeed. Once they demonstrate the economic value of their revolutionary technology, large, well-funded companies enter the market. The undercapitalized startups simply cannot compete.

To demonstrate the point, here are a couple of the more interesting examples of success and failure in two of the four technologies just mentioned. The startups that prospered had the timing right; the failures did not.

Transistors. The promise of the transistor was evident at its invention. In 1950, just 3 years later, a group of engineers and scientists started an entrepreneurial company, the Transitron Corporation, to manufacture germanium transistors. It didn't last long. Unable to compete with the fast-paced technology development programs of big competitors, Transitron went out of business in 1955, long before the transistor industry hit its stride.

Intel, on the other hand, was founded much later, in 1968, when silicon transistor technology was much better developed. Its timing was excellent, and the company went on to become an undisputed industry leader in ICs.

Semiconductor lasers attracted attention when their technology started to mature in the late 1960s, but no proven applications existed. In 1968, when small-scale military applications emerged, Laser Diode Laboratories was founded to produce the devices. The company faded into oblivion in the early 1990s, a casualty of its inability to keep up with the rapid technology improvements needed to fit lasers into the booming fiber optic communications industry.

Ortel Corporation, a much timelier startup, was a success. Founded in 1982 on the basis of technology developed at the California Institute of Technology, and partially funded by Warburg Pincus, Ortel entered the market just as fiber optic communications became commercially important. After a public stock offering on NASDAQ, Ortel was acquired by Lucent Technologies in 2000 for $2.95 billion.

The importance of synergy

Assessing the potential value of new technologies is much more difficult when you base your forecast of the future solely on the merits of an *individual* innovation. You have to take into account the synergy of several technologies working together. By contrast, startups with a narrow technology focus may very well end up bypassed and unable to benefit from synergies that emerge later.

This is because technological progress is never linear. Inventions can lie dormant for years without making a significant impact in the market. Then one day an apparently unrelated but complementary innovation appears, combines with a prior invention, and produces an unexpected and valuable result.

For example, low-loss glass fibers for transmitting light were invented in the late 1960s. Yet they would not have gained market success as the medium for fiber optic communication systems without the enormous progress in semiconductor lasers and computing that began around the same time.

Ironically, observers originally saw these three areas of innovation as unrelated. Lasers had no commercial applications, and in fact were regarded as the stuff of sci-fi stories about "death rays." Computers were associated with databases and calculating power, not telephone calls. While optical fiber was best suited to carrying digital information, the phone system was analog.

But together these three technologies made it possible to build high-speed digital fiber optic communication systems. Eventually they enabled the Internet.

Cautionary tales

Investing in companies with technologies that promise to obsolete incumbent market leaders is obviously tempting, and it can create great value if successful. We will recount one such success story in Chapter 5, where we discuss Level One Communications.

But mainstream technologies don't stand still either, as those who attempt to short-circuit their market dominance with radical new innovations sometimes discover too late. Just when you think that current technologies have reached the end of the road, mainstream players come up with innovations of their own to prolong their market reign.

Here are two examples to illustrate the point.

The first concerns a proposed breakthrough technology to enable supercomputers. In 1983 Warburg Pincus helped start a company to commercialize superconducting devices on the basis of a technology developed at the IBM research laboratories. These devices were intended to be the enablers of extremely powerful computers.

At that time, the performance of superconducting microwave devices could *theoretically* be expected to eventually exceed that of silicon ICs in supercomputing applications. But there were big hurdles in the way of practical implementations. First, the devices would have to operate at very low temperatures – close to that of liquid helium. This required special refrigerators. Second, and much more daunting, was the difficulty of manufacturing such devices with the complexity needed for powerful computers.

However, because of the potential of this technology, the company received the enthusiastic financial support of the National Security Agency and other Federal agencies that required special, very powerful computers for intelligence gathering and analysis. These contracts focused the company's efforts on solving the practical problems.

While the company made steady progress in doing so, the folks working on silicon chips did even better. Enormous investments by the mainstream semiconductor industry produced silicon IC devices so advanced that the anticipated advantages of superconducting devices in supercomputer applications disappeared. As a result, the original hope was frustrated and the project stopped.

But this was not the end of the story. In an example of how new technologies sometimes find valuable applications in unexpected ways, the company discovered a market niche it had not anticipated when the investment was made. In 2008, fully 25 years after its launch, the company continued to derive revenues from superconductivity research contracts funded by the Department of Defense. This is because superconducting microwave devices have unique applications not filled by other devices.

Of course, as a venture capital investment, this startup failed to produce an attractive return. It just took too long to build the business, and the company ended up much smaller than anticipated.

Our second example illustrates how unforeseen developments in competitive technologies can derail sober projections by industry experts. We wisely took a pass on funding a company to commercialize X-ray lithography, a technology that was supposed to replace the light-based lithography used in the manufacture of IC chips.

Photolithography uses light sources to generate device patterns on chips, much the way negatives are used to produce prints of pictures taken with film cameras. (Since digital photography has relegated the film camera to the status of an historical artifact, young people will have to consult Wikipedia on film technology to understand the previous sentence!) Photolithographic machines are very sophisticated, costing many millions of dollars.

Over the past three decades the dimensions of minimum device features on chips have shrunk from several microns to a small fraction of one micron. Pundits predicted that when they got much below one micron, they would be smaller than the wavelength of available high-power light sources. Then, it was predicted, chip makers would need a new generation of equipment that replaced light with other, shorter-wavelength sources of energy.

This appeared to offer an opportunity for businesses using radically new technology. In the 1980s and early 1990s we saw some proposals to build machines based on X-ray sources, ion beam sources, or electron beam sources. Each of those technologies had been explored in a laboratory environment, but none were yet commercialized for chip production.

At first glance this looked like a huge opportunity to build a major new equipment company. But on reflection it was not at all clear that light-based lithography was indeed doomed to reach its limits at the predicted dimensions. This was a tenuous basis on which to commit a large, possibly undefined, amount of funding to a startup.

Adding to the disincentive was the fact that going from a lab prototype to a production tool is such a huge step that it is hard to predict how much it will cost and how long it will take. We will have more to say on this topic below, when we consider types of technology risks. A new company would need years of expensive development work, followed by extensive customer trials, to build and market such untried products.

We turned down all investments in new silicon lithography machines for two reasons. First, it was not clear that a radically new technology was actually needed to build chips with submicron features; second, commercializing such products would be very difficult.

Any such investment would have been a bet against mainstream development work then being conducted by hundreds of scientists and engineers. While some experts were certain that light-based lithography was obsolescent, many others kept working on solutions to extend its life. In fact, we were aware that creative ideas were surfacing in the industry around the use of new laser light sources. There was enough doubt about who the eventual winners would be that an investment would have been unwise.

Subsequent events proved the wisdom of this decision. Betting against the majority view in the industry would have been disastrous. In due course ultra-violet laser light sources appeared that enabled chip features in the 0.020 micron range. A few large companies producing this equipment, including Canon and Nikon, are the same ones that have traditionally dominated the market for photolithographic machines for many years.

PROTECTING INTELLECTUAL PROPERTY

Any company that depends on proprietary technology for its competitive advantage knows that it must be vigilant in guarding its intellectual property (IP).

In an earlier era things were different. Proprietary manufacturing techniques served as sufficient obstacles to competition from knock-off products. But today many electronic products rely on software-based features rather than hardware to differentiate them from competitors. It is much easier for other companies to mimic their functionality, look, and feel.

A quick glance at non-Apple music players will demonstrate this truth. The clones look remarkably like iPods. They have not dented the iPod's market dominance as yet, in part because Apple's IP keeps them from duplicating its innovations.

It is not surprising, then, that technology companies are focusing on protecting their IP. They see owning patents as necessary for corporate survival.

Patents vs. innovations

How far this is true depends on the nature of the patents. While IP is obviously important, entrepreneurs frequently claim that their patents will provide a foolproof basis for launching a business. That claim is rarely sustainable, in spite of some famous examples, such as Xerography.

This is because the overwhelming majority of patents, while *legally* defined as inventions, are actually evolutionary developments of earlier work. The earlier work is frequently in the public domain because the patents on it have expired. There is nothing to stop other companies from basing competitive products on the same predecessor inventions.

As a business matter, then, it is crucial to keep in mind that patents do not confer a competitive edge in and of themselves. While patents on key aspects of a solution are valuable, responsiveness to market demands is more vital to a venture's success. It is more important that a company's product or service solves an important industrial or consumer problem in a new and greatly improved way than that it be based on patented technology.

Litigation hazards

Patent lawsuits are part of the competitive game. They are especially worrisome for startups. When a small company poses a threat to a market incumbent, it makes a particularly tempting target for this tactic, for reasons we outline below.

Ambiguity is the mother of these lawsuits. Since most patents build on what is called "prior art" (earlier inventions), few issued patents are clearly unique and isolated from the past. Courts are constantly being asked to arbitrate infringement based on arcane and frequently indiscernible differences between competing claims. As a result, companies frequently find themselves in prolonged, costly patent litigation, both offensive and defensive.

Furthermore, companies face patent suits from a new breed of businesses, denoted "patent trolls." These firms have no assets except for lawyers and acquired patents, and they make a living by threatening patent litigation in order to extract license fees.

Patent litigation is growing, along with the number of patents filed and issued. Figure 3.1 shows the number of patent applications filed in the US since 1963. Note the increased rate of filing after the 1980s, when it became possible to patent software – although such patents have been controversial.[2]

The growing level of patent litigation can be gauged from the number of patent suits settled. As Figure 3.2 shows, this figure more than doubled between 1988 and 2007.[3]

We have always had patent fights, of course, but these days the legal battles are more intense. Why has the pace of patent litigation increased so much? One reason is the rise of specialist product companies.

There was a time when the electronics industry was dominated by vertically integrated giants such as Siemens, RCA, GE, IBM, and AT&T. Each company maintained its competitive edge with strong patents and proprietary approaches to product design and manufacturing. For example, RCA, GE, and IBM produced their own semiconductor chips, which gave their equipment a marketplace advantage.

It was also common practice for the giants to cross-license their patents. This was not just to make life simple. Since their patent portfolios were so large, and these behemoths often needed to implement a patented technology for a new product, cross-licensing with rivals served to prevent endless litigation.

---

[2] B. Klemens, "Software patents don't compute: No clear boundary between math and software exists," *IEEE Spectrum* (July 2005), 56–59.

[3] Federal Judicial Center Research Division, *Annual Report of the Director of the Administrative Office of the US Courts*, Table C-4.

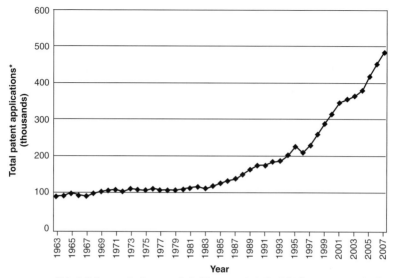

Note: * Reissue patents are excluded from counts in the "Total patent grants, foreign origin percent share" data column for the years 1963 to 1969; counts of applications exclude reissue patent applications.

FIGURE 3.1. Number of patents filed annually in the US between 1963 and 2007, excluding reissue patent applications. Around 50–60 percent of these are finally issued after about 33 months. *Source:* Based on data from US Patent and Trademark Office.

Industry giants were not as generous to small companies or serious competitors without any corresponding patents to cross-license. In those cases they derived substantial revenues from licensing patents that were not strategic to their own businesses.

Today, however, the days of easy licensing are long gone. The industry has fragmented into specialized vendors that depend on their IP as a competitive differentiator. Now companies with strong IP portfolios use their patents to keep competitors at bay. This is particularly true in the software industry, which has enjoyed the ability to patent programs since the 1980s.

All of these developments add up to a significant threat to startups in the form of a focus on IP issues on the part of large companies. Newcomers can find themselves facing serious patent litigation launched by entrenched market leaders. It is not uncommon to

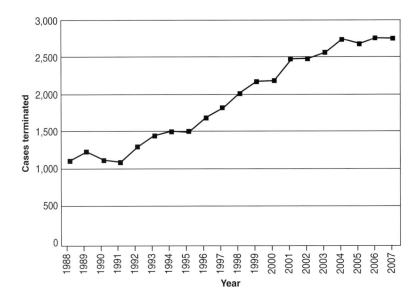

FIGURE 3.2. Federal district court patent lawsuits that were
terminated by fiscal year. Terminated cases include judgments,
dismissals, settlements, transfers, and remands. *Source:* US Federal
Judicial Center Research Division (ref. 3).

find that such litigation attacks are done not in good faith, but as a
method to harass a potential competitor.

It can take years of effort and millions of dollars to settle such
cases. During this time a new company can find itself distracted
from its main mission and on the hook for large legal costs. Although
there is no foolproof way for investors to guard against such risks, it
is common practice today to conduct thorough patent searches to
identify any potential conflicts prior to investing.

Companies in all stages of development also typically encourage their employees to file for patents. It is a cost well worth incurring to protect IP.

No patents required
In this context it is worth noting that a company can be built on
an important innovation without any patents at all. This is done by

combining common technologies in a clever new way to produce a valuable product or service.

One good example of a company built on a non-patentable innovation is Nova Corporation. We will present a detailed case study on Nova in Chapter 7, but in a different context.

The founders of Nova saw an opportunity to address a market that big players in the credit card processing business were ignoring: small merchants who wanted to accept credit card payments from their customers but could not justify paying the high access costs of the credit card networks.

They came up with a greatly improved way of communicating between the merchant terminal and the credit clearing infrastructure. When we analyzed the proposed business plan, we found that the technology would most likely work, but it was not patentable.

Nova eventually built a customer base of over 500,000 small merchants distributed across the US. The company succeeded not because of a firewall of patents protecting its innovation, but because of its ability to use clever marketing to quickly leverage its technology into a profitable business.

TECHNOLOGY-SPECIFIC RISKS

Worries about investment timeframes, market readiness, patent protection, and litigation threats confront the investors in virtually any company, no matter what its technology base. But there are other risks that are specific to individual technologies. We will now consider how these affect investment decisions in three major technology areas: software, ICs, and manufacturing.

Software development

Some companies sell software as a product. Others develop software for internal use, to enable new services or products they plan to offer. Both business models are dependent on the success of their software development efforts.

Software programs vary widely in their complexity. The more complex they are, the harder it is to complete them within predicted cost limits. The iron rule is to expect the development time to exceed forecasts and to cost more than planned – sometimes a lot more.

Another axiom about creating software is that the quality of the development team and its leadership are all-important. Anyone who claims that new methodologies and tools have made software writing a predictable mechanical task has never lived through the development of large programs. There are also enormous variations in the software programs produced by different teams even if they are working toward the same end specifications.[4]

Complex software programs can take years to complete and too often contain bugs that are discovered too late – after the program has been "fully tested" and released. Not even companies with vast programming resources have been able to avoid this problem. New operating system software from Apple and Microsoft is never perfect – PC users expect each version to be followed by bug fixes, patches, and service packs to correct faults that went undetected during development. While methodologies have been invented over the years to make writing and testing code easier, it is still very much a skill-based process. There is simply no replacement for a highly competent development team.

A development team starts with a talented architect whose job is to structure the program. This leader is supported by a group of coders who execute the architectural vision. But even they are not interchangeable: There is a distinct hierarchy of value among coders as well. It is an accepted truism within the art that software engineers vary widely in talent, with the most valuable being ten times more productive than the average. We can debate the exact degree of difference, but it is large.

---

[4] For an interesting study, see B. C. D. Anda, D. I. K. Sjoberg, and A. Mockus, "Variability and reproducibility in software engineering: A study of four companies that developed the same system," *IEEE Transactions on Software Engineering*, 35 (3) (May/June 2009), 407–429.

Another common problem plaguing software projects is the way requirements change after the program has been started. Users are constantly redefining their needs, necessitating the rewriting of code. These changes ripple through the architecture of the program, leading to errors and delays.

We have mentioned that software is the differentiating factor in most modern electronic products, so it is not surprising that new software plays a role in practically every investment. Given the software development hurdles we have just listed, how can an investor assess the likelihood of a company meeting its development schedules and achieving its business plan?

There is no simple answer. Again, the most important determinant of success is the team and its level of experience in completing similar projects. Projects need experienced technical managers, plus a high level of skill among those who define the requirements. Such people know how to pick developers, monitor their progress, and understand when changes are called for. Think of them as grand master chess players, because the best are so rare.

Our experience with Maxis (discussed in detail in Chapter 7) is illustrative. This company produced computer games, which are highly creative software products. The company's fortunes were based in large part on the talent of its visionary lead developer and one or two others who came up with the idea and major features for every game. These unique individuals could have been complemented by other game developers to build the finished game, but they would have been very hard to replace if they had decided to leave.

At SynQuest (also discussed in Chapter 7), which developed software products for managing a manufacturing supply chain, the emphasis was on building a multifunctional team around outstanding architects who understood customers' needs. The development team included dozens of software engineers, but the product's commercial value was determined by its architecture, which was created by a handful of unusually talented people.

Semiconductor integrated circuits

Designing an IC involves the successful integration of complex electronic functions into arrays of millions of transistors to achieve the desired system objective. It requires experienced project leaders with intimate knowledge of the system requirements of the target customers. These leaders direct teams of engineers with both hardware and software expertise. Errors in design are extremely costly to fix, so team excellence is a key requirement.

In 1987 we financed Level One Communications, a startup focused on ICs for data communications between computers. The company was pioneering chips to transmit high-speed digital information over ordinary twisted-pair copper telephone lines.

In order to succeed as a commercial product, these ICs had to be capable of sending and receiving megabits of information across thousands of meters. This meant designing novel electrical circuits to automatically correct errors and cancel out the spurious electric noise that is unavoidable over long, unshielded copper lines.

Two factors gave us the confidence to make this investment: the excellence of the design team and the management skills of the CEO, a gifted technologist and business person. The technical team included engineers with prior experience at major laboratories in communications devices, the physics of transistors, and systems. While they were experts in their fields, the big unknown was whether they would be able to integrate all the complex and sometimes conflicting requirements into a single chip that could be manufactured at low cost with standard production methods.

Previously such a complex set of functions would have required several ICs functioning together on a printed circuit board. The Level One approach aimed at producing a cheaper and much higher quality solution.

The company was successful in creating and marketing this product, and in the process generated a number of important patents that protected its key innovations. Over the years Level One produced several generations of increasingly complex chips, but timetables for new product launches were frequently delayed due to unforeseen problems in major design projects. Delays of 6 months (or more) were not uncommon, but delays are preferable to hasty designs with consequent errors.

Fortunately, competitors did not get to market any faster. There is considerable art in building such products. This is not something that can be quantified. It depends on a constant correlation and analysis of data by highly skilled engineers, and affects all companies in the field to a greater or lesser extent depending on the talent of their teams.

Level One became a major global vendor of the Ethernet conductivity chips used in all computers today. It, too, is the subject of a case study later in this book, in Chapter 5.

Process technologies: The long road from lab into production

Process industries present the investor with an altogether different technology risk profile. Companies in these industries depend less on design and engineering than on applying advanced materials science and high-volume manufacturing technology to their products. An example will serve to illustrate the issues in such investments.

We invested in a young and highly entrepreneurial materials company in 1986 which, among other products, had developed a ceramic powder to be added to aluminum. The powder strengthened the lightweight metal. This made aluminum a suitable replacement for steel components in automotive applications where a combination of low weight and high strength was mandatory.

The market opportunity was especially attractive due to the automotive industry's growing interest in new materials. The strategy called for the company to transition from being a manufacturer of low-volume specialty ceramics to a high-volume vendor of novel materials in a potentially large, high-growth market.

This plan required the company to scale up the manufacturing of its silicon carbide powder from a laboratory operation to a full-blown production plant. That was no simple undertaking. Off-the-shelf production equipment was not available. Therefore the company not only had to scale the process up to production volumes, it also had to design special equipment for producing tons of powder to exacting specifications at low cost.

Because of unforeseen problems, the company's progress in this scale-up process proved much slower than anticipated. Problems arose in production that had never been seen in the laboratory. As a result, the company was unable to meet production commitments fast enough to satisfy its customers' expectations.

In any high-volume, high-growth market, customers develop supply alternatives as a matter of self-protection. In this case, the customers turned to other ceramic materials. The substitutes were in some ways inferior, but they were still adequate to meet their requirements.

In the end our company continued as a small vendor of specialty ceramic materials for low-volume needs. While this was a viable business, it obviously did not create the value we had hoped for when we made our investment.

The moral of the story is that scaling up novel process technologies involves timetables and costs which are difficult to predict. We are seeing the same drama play out today in alternative energy technology companies, with the last act yet to be written.

Take the production of solar cells, for example. Many academics and entrepreneurs have proposed new materials and processes to produce cheaper and more efficient solar cells for energy conversion. Mass production of solar cells creates value for investors only if factories can be built that produce devices at the lowest possible competitive cost. Developing production technology and building a factory to produce the cells may require many millions of dollars.

Even with complete confidence in the team and a conviction that the company has great value-creation potential, investors had better be ready to accept delays and their associated costs. That's simply par for the course in a process-centric startup pioneering new materials.

PLANNING FOR PRICE DROPS

Anticipating technology obstacles may be easier than anticipating the vagaries of the market. However, investors have little doubt that they will experience one market phenomenon: declining prices for their company's products.

Pricing is one of the "four Ps" of classic marketing theory, along with product, place, and promotion. How a company prices its product has a huge influence on the chances of its success in the market.

Marketers use a couple of established methodologies for setting an appropriate price for a product. One is to add up what it cost to design, develop, manufacture, and market it, and then tack on a reasonable level of profit. Another involves "value pricing," basing the selling price on what a typical customer might be willing to pay for the benefits conferred by the product.

These methods are valuable when dealing with a product line extension, or even a new product in an established market. But when a company introduces a breakthrough product that creates a totally new product category, and is also facing a virgin market, pricing becomes more art than science.

Instead of comparing the offering to other, similar products, the marketer has to ask such basic questions as, who are the customers? What do they want? What are they willing to pay? It is hard to quantify the market opportunity when you are dealing with intangibles like these.

Once again, look at the example of Apple's iPod. Few people remember that in 2001, when the original iPod appeared, there were already several hard disk-based portable music players on the market. At its introduction most consumer electronics experts pronounced it grossly overpriced at $399, at least in comparison to its competition.

Were they ever wrong! The iPod simply outclassed the existing players and took over the market, in spite of its premium price. Apple had priced the iPod just right for its perceived value, and reaped the rewards with a highly profitable product line.

But that's not the end of the story. There is another pricing phenomenon that affects most new technology products, and which VCs ignore at their peril. It is the general inability of electronic products to hold their prices.

Let's look at Apple again. At the time of this writing, Apple's "iPod classic" still sells at a premium, and continues to dominate the market. Yet its selling price is only $249, about 60 percent of the money it commanded at introduction. However, buyers get more product, not less, in spite of that 40 percent or more discount. The Classic now has a storage capacity of 120GB, 40X that of the original model, and can handle photos and video as well as music. Apple has also broadened its line with other models that sell for as little as $49.

Far from being extraordinary, the price drop on the iPod represents the norm for an electronic product. Price erosion in the technology sphere is not a risk, it is a certainty. Broadly speaking, the unit prices of technology products have declined sharply along historical experience curves. Several, like the iPod, sell at only 60 percent of their original price levels within 6 years of introduction.

Research has confirmed the need for a business plan to anticipate the inevitable declines in product prices, and project ways to reduce costs in response. Gottfredson, Schaubert, and Saenz, for example, have studied historical pricing trends in many industries.[5] They trace historical trends in price reductions that point out an interesting relationship between prices and production volumes. As sales of electronic products rose, and more efficient manufacturing technologies were implemented to meet demand, prices dropped rapidly.

---

[5] From M. Gottfredson, S. Schaubert, and H. Saenz, "The new leader's guide to diagnosing the business," *Harvard Business Review* (February 2008), 1–12.

For example, between 1980 and 2005, every time production volume increased by 60%, the average unit prices of microprocessors dropped by 40%. A similar trend is apparent for liquid crystal displays between 1997 and 2003.

In effect, production costs and selling prices in technology industries move in tandem. More efficient manufacturing makes it less expensive to make products, while competition forces companies to pass the savings along to the consumer. Companies hoping to prosper in fast-growth industries have to stay on the cost curve or disappear.

COMPRESSED LIFECYCLES, FASTER PRODUCT DEVELOPMENT

Risk assessment is not limited to technology issues or pricing trends. It must also take accelerating obsolescence into account.

Ask yourself, who buys vinyl LPs today? Music on cassettes? Movies on VHS videotapes? Yet these products were ubiquitous as recently as 20 years ago.

During the past few decades technologies have become obsolete faster and faster, and, as a result, product lifecycles have become much shorter. In consumer electronics, product life cycles are in the 18-month range.

The chart in Figure 3.3 tracks the market penetration of various popular consumer products since 1947, starting with black and white TVs and moving to digital TVs in the 1990s and 2000s.[6]

Obviously the rate of product acceptance has accelerated over the years. This indicates that manufacturers are increasingly willing to introduce new products that obsolete old ones at a fast pace. Note how the products with the steepest curves – DVD players and digital TVs – are the most recent. Market acceptance

---

[6] H. Kressel with T. V. Lento, *Competing for the future: How digital innovations are changing the world* (Cambridge: Cambridge University Press, 2007), p. 302. Reprinted with permission of Cambridge University Press.

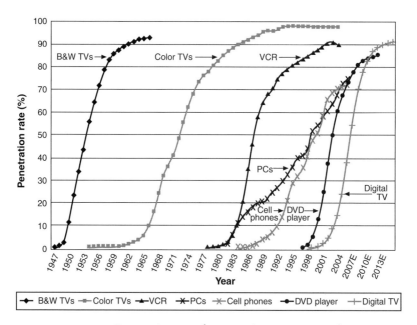

FIGURE 3.3. Penetration rate of some major consumer products in the US since 1947. *Source:* Based on data from IDC and Consumer Electronics Association (CEA), *Market Research* 05/07 (ref. 6).

happened faster with each new generation. And in most cases these newcomers totally displaced existing product categories.

Accelerated obsolescence of major technologies is mirrored by the continuing compression of product lifecycles. Where companies like Apple, Nokia, Samsung, and Sony once introduced major products a couple of times a year, they now bring out new items and whole new product lines on an almost continuous basis.

It doesn't stop there. Rapid obsolescence of end products translates into equally rapid obsolescence of the components that go into those products. This affects software, electronic devices, and systems alike – favorite areas for venture capital investment. Faster product development has become a necessity throughout the supply chain of the electronics industry.

Consumers have embraced this constant barrage of innovative gadgetry. Telephones, for example, lasted for decades in the days of Ma Bell. Now users trade in their mobile handsets for newer models – and their mobile carriers for providers with better phones – after less than 2 years of ownership on average. That is a rate of obsolescence more typical of the fashion industry than the traditional equipment market.

The twenty-first century phenomenon of fast turnover in product lines is driven by changes in features and functions. Just 10 years ago, for example, the nature of an electronic product was determined by its hardware, its physical construction. To make significant changes in a product, you had to rework not just your concept, but your whole manufacturing supply chain. That was expensive, time consuming, and a major disincentive to frequent updates of features, functions, or design.

As long as hardware dictated the nature of a product, product lifecycles extended for years. Two relatively recent developments have conspired to overturn this long-accepted situation: the increasing importance of software, and ever more powerful, commercially available ICs.

Same chips, different products

If you remove the cover of practically any electronic product you will see one or more circuit boards stuffed with IC chips and other components. Most, if not all, of them are made by third-party suppliers. When you open a competing product, you will see some of those same chips on its boards, too.

Chips from Silicon Optix, for example, do the video decoding in dozens of home theater receivers from various competing manufacturers. Digital signal processors from Texas Instruments show up in audio, video, communications, computer, and industrial products from many different companies. Intel microprocessors are the brains inside an endless number of PCs and other products. No matter what brand of mobile phone you use, the chances are that its power amplifier, which transmits the signal to the receiving tower, comes from one of three major makers.

Yet while electronic products increasingly use the same standard, commercially available chips to perform basic functions, they are not clones of each other. Embedded software allows designers to customize their features, functions, and user interfaces to differentiate them from competing products.

In essence, many of today's electronic products incorporate specialized computer systems in which proprietary software, not hardware, is the key driver. They function as "flexible product" platforms.

Companies that make the costly up-front investment to create such product platforms can quickly alter them to meet developing market trends. If you develop new software, and perhaps revamp the external styling, you have a new product, defined by its new features. As a result, equipment suppliers regard innovative software talent as a key competitive differentiator. It represents their biggest investment in product development.

For the typically cash-starved startup, this is a serious problem. How can it cope with the accelerating demand for new products? It is faced with the necessity of churning out one new offering after another just to stay in the game.

Product development consumes a huge part of a startup's capital. Startups have to keep designs fresh or risk falling behind. With overnight obsolescence and fast model turnover being the norm in the market, new companies cannot count on riding a single product to long-term success. They have to find a way to introduce new products at frequent intervals while generating good profits from existing products.

That is why practically all electronic products companies minimize infrastructure investments by concentrating on product development and marketing, and outsourcing manufacturing to contract production companies.

TECHNOLOGY COMPANIES GO GLOBAL

The growth of international markets is another key factor in most investment decisions.

Globalization in this sense started in the 1980s with the massive migration of manufacturing to Asia. Product engineering followed a few years later. The result has been the transformation of world trade patterns, blurring the old division between "industrialized" and "developing" nations.

Figure 3.4 shows the strength of this trend.[7] It compares the distribution of trade in high-technology products among various regions in 1980 versus 2005. Asian economies (outside of Japan) have increased their share from only 8% in 1980 to 37% in 2005. Over the same time period the US share has declined from 29% to 12%, and that of the European Union from 40% to 28%. But of course the pie has gotten bigger: world trade increased from $159 billion to $2.26 trillion, so everyone benefited.

Ambitious entrepreneurs understand the implications of this tidal shift. It is no longer enough to operate exclusively within your own country. If you want to build a bigger business, you must aspire to a global presence.

While the dispersal of economic activity has spread the risk of market entry for startups, it has not made things easy for the investor. Not only is it more difficult for the investor to forecast results, it is much harder for an actively involved VC to keep an eye on business operations in a global enterprise than it is in a national or regional company.

Globalization complicates company management, too. Building a business when people are spread across many time zones is no easy task, and neither is compensating for cultural differences.

---

[7] From "Measuring the moment: Innovation, national security, and economic competitiveness," *Benchmarks of our innovation future II: The report of the task force on the future of American innovation* (November 2006), p. 14 (accessed October 3, 2008 at http://futureofinnovation.org/PDF/BII-FINAL-HighRes-11–14–06_nocover.pdf).

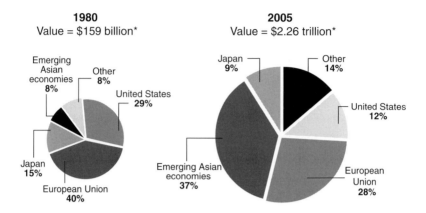

Note: * 2000 US dollars. Emerging Asian economies: China, South Korea, Taiwan, Singapore, Hong Kong, India. High-tech includes aircraft, pharmaceuticals, office and computing machinery, communications equipment, medical, precision, and optical instruments.

FIGURE 3.4. The fraction of world trade in high-tech products in 1980 and 2005 by region of origin. The total value of this trade is indicated. *Source:* From www.futureofinnovation.org, with updates of the 2005 data from the American Physical Society (ref. 7).

Team-building activities and managerial oversight are both bound to be more challenging in an international company.

Operating across geographies

To globalize a business, you must learn to distribute resources across all the geographies where you sell products or conduct other activities.

For "hard" product companies this means finding cost-effective and high-quality manufacturing sites. Service companies need to set up local branches where the infrastructure is modern and the work force large and educated enough to support operations.

In an effort to maintain its competitive edge a global company may choose to complement its own resources by using contractors in every geographical market where it operates. For example, a company can either outsource its customer support functions or maintain internal staff for that purpose, depending on the needs of the particular geography. The issue is not just who the customers are and what they want. It is also where they are and how you can meet their needs.

We have already pointed to declining prices as a way of life. In a global enterprise price becomes even more crucial, as its products are sold into low-income countries, not just wealthier nations. The company's success in less affluent markets depends on its ability to get product prices down to where local buyers can afford them.

An obvious strategy for reducing costs is to set up engineering centers in low-cost developing nations with pools of well-educated engineering talent. This helps keep product prices competitive in the developing world, and it also cushions the blow when the inevitable process of price erosion begins.

Global from the get-go

RMI Corporation, funded by Warburg Pincus and others in 2003, demonstrates how startups can create a global footprint right from the beginning. RMI designs and markets highly sophisticated semiconductor chips for network and video processing. Its customers are in Asia, the US, and Europe.

To serve this widely dispersed customer base, the company has located some of its functions in different sites based on available resources and local needs. It located its product design in California, to take advantage of an extraordinary pool of local talent that makes it still the best place to do this kind of work.

The software needed for product delivery, however, is written in India and China, where costs are much lower. Product localization and customer support are done in several different locations. The whole organization is closely linked with its California headquarters.

While an international stance is no guarantee of success, globalization gives a startup company more options for sourcing technology and building sales. In many ways RMI is a paradigm of the modern technology company. Its management staked its future on being a global enterprise. By tapping the best and most cost-efficient sources of innovation and management skills wherever they could be found, it positioned its products for sale into the world market.

INDUSTRY CONSOLIDATION: A WAY OF LIFE

We come at last to a fact of commercial life that impacts the investment required to fund a startup, and also determines its ultimate fate: industry consolidation.

Consolidation is a normal process. Eventually every major industry comes to be dominated by a handful of large companies. This helps explain why most startups, rather than survive as independent companies, are acquired after a few years of separate existence.

Consider the following examples.

- How many companies still produce small computers? The dozens of players that competed for market share in the 1980s, including such once-formidable names as Compaq, DEC, and Wang, have shrunk to about half a dozen significant vendors. Not one of the three companies just named is among them.
- How many companies are fighting for dominance in mobile phone handsets? Only five brands now account for over 80 percent of the global market: Nokia, Motorola, Samsung, Sony Ericsson, and LG.
- How far has consolidation gone in semiconductors? The top ten companies represented 44% of global revenues in 1996. Ten years later they accounted for 56%.

These indicators just scratch the surface. Look at communications equipment, consumer electronics, and industrial products of all kinds, and you will see the same pattern. New leaders emerge and older ones disappear.

It is not just small companies, either. An interesting perspective on large, publicly held companies is shown in Figure 3.5, which plots the average tenure of companies on the S&P 500 list since 1928.[8]

---

[8] This figure appears in R. Foster and S. Kaplan, *Creative destruction: Why companies that are built to last underperform the market – and how to successfully transform them* (New York: Doubleday, 2001), p. 13.

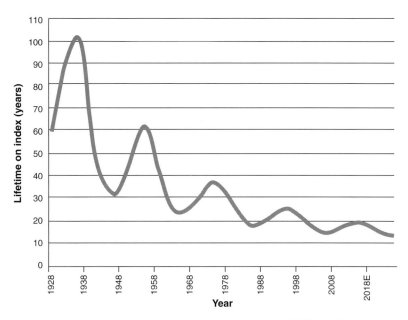

FIGURE 3.5. Average tenure of companies on the S&P 500 list since 1928. *Source:* Reprinted with permission from McKinsey & Company (ref. 8).

As the chart shows, the average tenure on this list is now only 15 to 20 years. Companies disappear from the list through bankruptcy, mergers, or a relative reduction in their market cap as other firms overtake them. To understand how this can happen to such powerful entities, note that in 2003 the top 25 US corporations in the technology sector purchased over 500 other companies.

The big get bigger

While the lifetime of big public companies might be limited to decades, that of young companies is even shorter. This is actually beneficial for VCs invested in good companies, because when fast-growing businesses with outstanding products and unique technologies are quickly acquired by industry consolidators, investors get their investment back faster.

In a classic absorption process, companies that acquire startups might be acquired in turn by even bigger companies, forming industry sectors with a small number of dominant players. How long even those market leaders stay on top, however, is unpredictable. New technologies and market forces are always threatening to turn industries upside down.

What drives consolidation? Here are the major factors, some of which we have already touched upon.

- Most technology product companies have to operate globally. This favors large businesses. Having operations dispersed around the world means additional overhead. Companies need large revenues to support this overhead and still sustain profitable operations.
- Technology-based companies find it imperative to acquire companies with complementary products to grow their product portfolios. Acquisitions require access to large amounts of capital, again favoring the big companies.
- Innovative IP is essential to continued success in a constantly changing market. Large enterprises can build a large portfolio of patents and trade secrets more easily through acquisition, helping them resist attacks by competitors.
- As industries consolidate, suppliers tend to follow suit. Big companies are most comfortable dealing with other big companies as suppliers of critical products and services.
- Brands are important drivers of global distribution and customer acceptance of new products. Companies with strong brands, such as IBM and Hewlett-Packard, have a major global competitive advantage. They can leverage their strengths by buying companies with good technology but much less marketing power.

In the face of the inevitability of consolidation, investors must decide on an exit strategy which will be gated largely by the availability of capital to finance growth, balanced against the company's assets and management talent. Unless the company can hope to become an industry leader on its own, which usually involves becoming a public company at some point, being acquired is a good way for investors to realize the value they have built up while their products and technology are still relevant.

More on consolidation: Software companies bulk up

The software industry continues to attract many startups. But as fast as young companies appear, they are acquired by bigger companies searching for new products to bring to market without doing internal development.

Figure 3.6 shows how consolidation has changed the landscape of the software industry. Between 1996 and 2007 the number of medium-sized software companies grew dramatically as vendors competed to gain critical mass in the market. They achieved this goal mostly by acquiring other companies.

Another big motivator for growing the revenues of a software company is to improve its profitability. Scale matters in generating profits. In 2007 the ten biggest players accounted for 68% of the revenues of the industry and fully 90% of the profits. This leaves little for the small guys competing in a consolidating market sector.

Small software companies have a hard time remaining profitable in the face of relentless competition because they are forced to keep increasing their investments in product development and marketing just to stay in the game.

DUE DILIGENCE

We return to where we started. In any investment in a young company, the ultimate objective is to assess the probability that the company under consideration will be worth substantially more than the capital invested in building it.

Such an assessment assigns relative weights to many of the elements needed for success, and balances them against the risks the company faces. It is not a scientific process. The result, more often than not, is an investment decision that is basically an informed guess, based on experience and the careful analysis of available facts.

Does this all sound terribly risky? It is. But successful investors diversify their bets and learn to improve their chances of success by

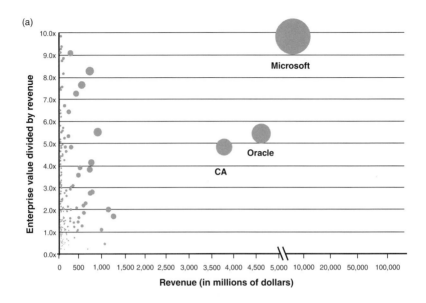

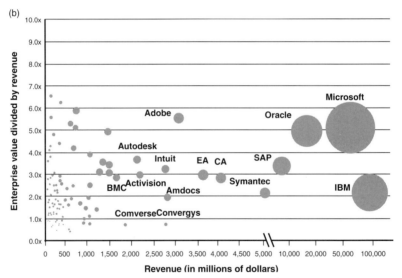

FIGURE 3.6. Enterprise value divided by revenues of the largest public software companies as a function of annual revenues in (a) 1996 and (b) 2007. *Source:* Based on data from Capital IQ.

using a portfolio strategy. They also have access to expert talent and an investment industry network.

Most important, they have experience. Over the course of time they develop a sense of how to balance risks and rewards. They know that if you are too conservative, nothing gets done. Be too aggressive and you wind up with a disastrous portfolio.

Throughout this chapter we have focused on the considerations faced by venture capital investors in making their investment decisions, and on the outside forces that affect a company's fortunes once an investment is made. In the next chapter we start our review of thirteen investments and the lessons learned from them.

4 Investing in a transformed market: Telecommunications

> Light is an old friend ... The visionaries who foresaw a wired city were wrong – we will have a fibered society instead. We can all watch it happen.[1]

Communications of one kind or another have always played an important role in human affairs. Battles have been won or lost, tragedies averted or assured, empires strengthened or imperiled by the timely delivery of crucial messages. Yet for most of recorded history the technology of communications remained static.

The first big leap forward in communications technology did not happen until the nineteenth century. That is when the invention of the telegraph and the telephone ushered in the age of electronic communications.

It took another one hundred years for the next dramatic advances to arrive. And they were giant steps indeed. Satellite communications was one major new invention. But the most important advance of all, digital technology, first appeared commercially in the 1970s. It quickly displaced analog systems, revolutionizing the way people around the world communicated with each other. Digital communications also transformed the world economy.

The invention and deployment of fiber optic systems, electronic switching, and other technologies associated with digital communications was fueled by multi-trillion dollar investments. Some of them came from venture capital funds, including Warburg Pincus.

In this chapter we will discuss five investments Warburg Pincus made in the telecommunications field before 2000. These were selected to illustrate how investments emerged in an explosive

---

[1] J. Hecht, *Understanding fiber optics*, third edition (Upper Saddle River, NJ: Prentice Hall, 1999), p. 2.

growth industry where multiple avenues to build valuable companies appeared to exist. We will discuss the investment thesis for each, review what actually happened, explore the reasons for their success or failure, and discuss lessons learned.

And, yes, there were two disappointing investments among the five companies selected. The telecommunications markets that emerged out of the 1970s promised huge opportunities for new businesses, then delivered its allotted portion of disappointments. We will discuss our qualified failures because their history offers some valuable lessons.

As our five examples will show, there is no such thing as a sure formula for success, even in a wide-open new market. Better technology, financial resources, market position, structural upheaval in the industry, changing product standards – these and many other factors affect the fortunes of companies. Most critical of all, perhaps, is good timing, trumped only by the quality of management. And a little luck is always welcome.

When I explain to people the role chance plays in the success or failure of investments, I am met with disbelief. Aren't professional investors supposed to have superior wisdom? I remind folks that even among the professional prophets of myth and history, only Cassandra was right every time – and no one believed her. In the business world, we have only amateur prophets, and they have very spotty records.

That is what makes venture capital investing so fascinating. Although reliable prophecy might be unavailable, experience and acquired judgment count for a great deal. Many of the issues that arise when making an investment in one technology company are sure to reappear in one form or another in other investments. That is why focusing on lessons learned is such a useful exercise.

Five companies, five different investment theses, and five different outcomes. To better understand how this could happen, we will preface our case histories with a review of the market landscape in which these businesses operated.

TELECOMMUNICATIONS TRANSFORMED

Look at today's teenagers with iPhones seemingly grafted onto their ears. They can't imagine being without a mobile phone. And they have never been restricted in their choice of service providers.

But there was a time when only top executives and government officials had mobile phones. They got them from AT&T, the government-regulated monopoly that dominated the US telecommunications industry. For most Americans AT&T was the only choice they had for any kind of phone service, and the company would not allow customers to use any equipment that it did not manufacture. The iPhone would have been locked out.

Our newfound freedom of choice is a direct outgrowth of a transformation of the telecommunications industry that took place over the course of only 20 years. It was a sea change in a once-stable market, spurred by radical new technologies introduced in the 1970s and 1980s, and enabled by deregulation in the 1980s and 1990s.

By the time it was over, the national monopoly was no more. The old, monolithic AT&T was divided up. Finally the last piece of the original company was bought by one of its corporate offspring, which promptly renamed itself AT&T.

Where AT&T once made all phone equipment, the slick and popular iPhone is made by Apple, a third-party hardware provider. Ironically, under an exclusive arrangement with Apple, at this writing the iPhone is available only through the new AT&T. This gives the carrier a huge market advantage, but there is no guarantee that AT&T or any other carrier will continue to have exclusive rights to the coveted device.

We funded the five companies under discussion here just as deregulation was turning telecommunications into a competitive battleground. Two of them, Licom and Epitaxx, were product companies. The others, NeuStar, Covad Communications, and IGS, were service providers. Each had a business plan for using new technology to satisfy unmet needs in the recently deregulated industry. Only a few years earlier starting such companies in these industries would have been out of the question.

ROOTS OF DEREGULATION

To put things in perspective, consider how different the market was prior to deregulation. A bit of history is of more than academic interest, because it explains the market forces we had to take into account when we funded our companies.

Telecommunications was regarded as a tried-and-true industry with a century-long track record of stability and success. Its initial ascendancy had been swift. Within a few years of Bell's original invention, the telephone had established itself as a boon to modern life, and was considered a strategically important resource by countries around the world. By 1890 it had assumed such a large place in the popular imagination that Vienna's "Waltz King," Johann Strauss II, celebrated it in his Op. 439, the polka "By Telephone."

So important was the telecommunications industry that in most countries service was provided by government agencies or government-regulated private monopolies. In the US the original AT&T assumed that role. AT&T's customer pricing was negotiated with regulatory agencies, with the two sides working to ensure the company's profitability and the availability of universal service. Its stock paid such reliable dividends it was considered a "widows and orphans" investment – that is, a safe source of income.

By the 1980s, however, emerging new technologies were giving upstarts the tools to chip away at the edges of the telecommunications monopolies. Point-to-point wireless links, for example, threatened AT&T's dominance by partially bypassing its landline long-distance network. Digital communications over fiber optics promised a dramatic expansion in communications capabilities. AT&T faced aggressive new competitors who engaged it in acrimonious legal battles to force the old monopoly to cede some of its territory.

Eventually the whole idea of a monopoly was called into question. The regulated entity was no longer delivering the best technology. It was not serving the public with the lowest prices. It was time for a change.

Deregulation began with the 1984 breakup of AT&T into regulated and unregulated providers. Seven regional Bell operating companies (RBOCs) provided local connections under government regulation. The deregulated entity, still called AT&T, included a long-distance division and an equipment and software arm (later spun out as Lucent Technologies and what eventually became Telcordia Technologies, respectively).

AT&T quickly found itself facing fierce competition from new long-distance companies. In fact, Warburg Pincus financed two long-distance phone companies which became publicly traded: LCI International, funded in 1988 (acquired by Qwest Communications in 1999), and Trescom International, funded in 1994 (acquired by Primus Telecommunications Group in 1998). These successfully competed with AT&T for long-distance telephone service.

However, while long-distance telephony was competitive, local phone service was not. The RBOCs still owned the local exchanges and the copper lines that served the vast majority of customers. Companies that wanted to compete in local service faced the impossible chore of replicating the huge US telephone network.

Then came the Big Bang that changed the industry forever. The US Telecommunications Act of 1996 forced the RBOCs to rent access lines to the newcomers at fixed rates. It also mandated local number portability, permitting a customer to keep the same telephone number when switching to a new service provider.

Number portability was the Federal Communications Commission's (FCC) way of leveling the competitive playing field. No longer did customers who were comparing providers have to weigh the advantage of lower prices against the disruption of losing their existing telephone number.

It took 12 years (1984–1996) for the American telecommunications industry to complete its transition from a monopoly structure to open competition. The process created a totally new market, offering a potentially huge upside to companies that could deliver innovative products and services to the industry and its customers.

The situation was tailor-made for entrepreneurs and their comrades-in-arms, venture capitalists. Suddenly there was a new, deregulated market in an industry that was familiar to inventors and investors alike. Opportunities seemed to grow on every telephone pole. What more incentive did one need for starting a new venture?

Technological advances also made life interesting for entrepreneurs and VCs during this period. These, of course, grew out of the switch from analog to digital transmission. However, the full power of digital technology could not have been realized without optical communications over glass fibers.

FIBER OPTIC DIGITAL COMMUNICATIONS

Fiber optic cable is the foundation of the digital revolution in telecommunications. The extraordinary increase in bandwidth that it enabled provided the platform for explosive growth in the industry, both technological and commercial.

Analog telecommunications used electrical signals to carry content over copper wires from one place to another. Its basic principle, which had survived intact since the days of Alexander Graham Bell, varied the amplitude of an electronic wave form to correspond with the sound of speech.

Digital communications technology, developed in the 1970s, represented a new paradigm: the content was digitized and the data were transported and processed in the form of bits, either 0 or 1. At first the telephone companies used copper wires to carry digital data as well. To put the bits into a form suitable for the phone lines, digital communications used sophisticated coding of sine waves to represent the sequences of 0s and 1s.

As frequently happens with new technologies, it was the synergy between digital coding and fiber optics that sparked the explosive growth within the industry. It turned out that digital communications were ideal for transmission over optical fibers.

In digital communications the on–off switching of an electronic device produced the bits needed to transmit information. The resulting electrical signals were converted to light pulses by a semiconductor laser. These light pulses could be transmitted over glass fibers as thin as a hair, and reconverted into electrical signals at the receiving end.

Replacing copper or coaxial cable lines with thin glass fibers brought a huge benefit: it boosted capacity for high-speed data transport by many orders of magnitude. The distance over which links could be maintained without repeaters also increased to hundreds of miles. All of this drove a steep drop in transmission costs.

Commercial deployment started in a serious way in the early 1980s. By then many of the problems of optical communications had been solved, including the cost of the fiber, methods for bundling many fibers into a cable, the reliability of the photonic and electronic components, and an overall architecture that would allow practical systems to be built.

Although fiber links started as a backbone technology, they were eventually extended to reach business premises and even homes. Their high-bandwidth connections, impossible with earlier technologies, made the Internet not just feasible, but a daily reality in our lives.

Once deployment started, there was no going back. The potential impact of this revolutionary technology was too immense to ignore. Figure 4.1 shows the historical evolution of the information-carrying capacity in the US.[2] Until the advent of optical fiber technology, communication satellites and coaxial cable were

---

[2] C. Kumar and N. Patel, "Lasers in communications and information processing," in J. H. Ausubel and H. D. Langford (eds.), *Lasers: Invention to applications* (Washington, DC: National Academies Press, 1987), p. 47.

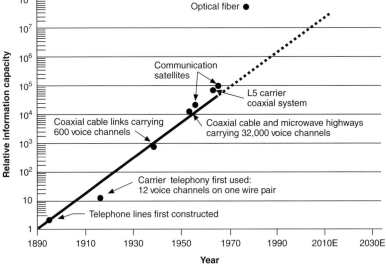

FIGURE 4.1. The growth of the relative information transmission capacity in the US since the 1890s. Note the huge jump resulting from optical fiber systems. *Source:* Kumar and Patel (1987). Reprinted with permission. Copyright © 1987 by the National Academy of Sciences, courtesy of the National Academies Press, Washington, DC (ref. 2).

expected to be the means whereby long-distance information would be transmitted. With the introduction of optical fiber communications, data transmission capacity went from being finite to becoming effectively infinite, as each fiber could be expected to carry gigabits per second over many miles.

Practically everything in fiber optic systems was new. On the photonics side the innovations included optical fiber manufacturing, cabling technology, optical light sources (semiconductor lasers), and optical detectors. In the electronics arena a whole new class of specialized processing computers was needed to encode and decode the transmitted data packets so that users could make use of their content.

As a result, there was a huge need for new components, modules, and systems that were not being provided by existing vendors. Entrepreneurs saw opportunities to build businesses in a virgin field.

The five companies we discuss here all took advantage of this technological sea-change in one way or another. We will look at two product companies first.

PRODUCT STARTUPS

In 1984 we were approached by groups of entrepreneurs looking for funding in product areas related to optical communications. When we considered the emerging business opportunities for product companies, it was clear that our selection had to be based on six factors:

- the qualifications of the entrepreneurial team;
- the non-infringement of intellectual property rights held by others;
- the probability of acceptance by customers;
- the size and nature of the potential market;
- the ability to sustain the company's competitive position against new entrants into the market;
- the scale of investment needed to get a product to market and finance its growth.

The product business ideas fell into three basic categories.

- *Companies making optical fiber and related products (such as cabling).* This was not an attractive choice. Corning Glass and AT&T had strong patent portfolios in the technology. Also, copper cable makers were already developing techniques for optical fiber cables.
- *Companies making electronic or optical components.* These included semiconductor lasers to transmit the packets, and light sensors to detect the optical signals and convert them back into electronic form. We already had one investment in semiconductor lasers, Ortel Corporation (mentioned in Chapter 3). Furthermore, large, well-known suppliers were also in the laser market and jockeying for competitive advantage. We considered that a newcomer could succeed with less glamorous devices, such as light sensors.
- *Companies manufacturing special data processing systems – in effect, special computers that managed voice and data traffic.* Since several large, established equipment vendors were already promising to sell electronic equipment that addressed the emerging needs of the industry, the challenge was to find a product that could beat them to market.

The product companies that we will discuss were founded in the early 1980s, just as the industry began serious deployment of the technology. One was Licom, a startup that developed an electronic system used to manage digital voice and data traffic. The other, Epitaxx, focused on manufacturing optoelectronic semiconductor sensors for fiber optic systems.

Licom: Betting on a new standard

The founders of Licom were engineers who had done pioneering work at large companies on telecommunication systems. Their proposed new product would transport data streams in packets based on a standard called SYNTRAN (from Synchronous Transmission), which was then emerging from industry standards groups. Their customers would be the telecommunications companies, primarily RBOCs recently liberated from under the old AT&T umbrella.

Our investment thesis was that Licom could become an industry leader by creating sophisticated electronic products that the newly deregulated RBOCs needed to build up their fiber optic networks. We estimated that the company could achieve $100 million in annual revenues once its products passed the acceptance tests in the industry.

In order to achieve success, Licom would have to get its products to market before the major equipment vendors could respond. Speed of execution was crucial, because we knew that the dominant equipment suppliers to the RBOCs, such as AT&T and Nortel, would eventually introduce competitive products of their own.

Simply entering the equipment business against big competitors was risky, so the company's strategy was to leverage a new technology as the basis of a product that was not available from other vendors. The investment thesis assumed that a newcomer with an early start and a unique product could successfully establish itself as a preferred vendor. That first product offering would then generate the revenues and cash flow needed to finance new products, keeping Licom ahead of the competition.

Licom opened for business in 1984 with an excellent team of technologists. The CEO had never run a business, but he had managed a communications system department in a large US government agency. Within 2 years of launch they delivered a product for testing by the RBOCs.

Product changes were needed over time, with the delays endemic to development processes that involve a great deal of software, as we learned in Chapter 3. Eventually the product worked as planned. But sales never materialized after the test stage, except for a small number of units delivered to U.S. West. Why? Because Licom was indeed first to market with a useful and unique product based on a new standard, but the actual standard never gained acceptance by the carrier customers.

SYNTRAN had been accepted by the industry standards-setting body as a formal standard to ensure the interoperability of traffic among all carriers. Yet there was a general view that this standard would eventually have to change to allow the management of higher data rates. The question was how long it would take for a subsequent standard to get general approval, so providers could start building products around it.

As interest grew in establishing the new standard, the standards-setting committee turned into an industry battlefield. Established equipment vendors worked hard to supersede SYNTRAN with a new standard called SONET (Synchronous Optical Networks), which was more comprehensive and offered an extended range of data transmission speeds. As the only vendor with a product featuring SYNTRAN, Licom was a lonely pioneer with no friends on the committee.

While SONET used similar concepts for handling the data packets, SONET-compliant products would not interoperate with SYNTRAN products. Watching all this standards-setting activity, the RBOCs decided to wait for SONET equipment to emerge from their established vendors instead of buying SYNTRAN products from Licom.

To make matters worse, this all happened in slow motion, while the company burned cash without significant revenues. The carriers kept enticing Licom with promises of purchases. They continued to test the product in the meantime, in order to become familiar with synchronous transmission technology. In effect, they used Licom's expertise to research the new technology at little cost to themselves – but unfortunately at great cost to Licom and its investors.

Even though U.S. West did buy some equipment, it was not enough to sustain Licom. A search for other markets for the product or its technology proved fruitless.

There was one option still open. Licom's investors could have decided to further fund the company to develop SONET products. However, the company ran out of money before the new standard was in place. Since it was therefore impossible to predict how much it would cost to field a new product, we decided against additional investment.

Because of the excellence of its engineering staff, Licom was eventually acquired by a large equipment company for a modest sum. That was in 1989. There was no return to the investors.

It should be noted that eventually SONET became the industry standard. It is in use to this day.

Epitaxx: Keeping it simple

In contrast to Licom's unfortunate history, Epitaxx successfully produced light sensors for fiber optic communications and turned into a valuable business.

Optical communications systems require light sources to send signals and light detectors to receive them. Lasers, which generate the light, are far more complex devices to design and manufacture than light sensors.

Not surprisingly, the optical communications industry put far more development effort into light sources than into detectors. Semiconductor laser diode technology was the object of huge R&D programs at major corporations around the globe.

Detectors or sensors, on the other hand, received much less attention, in large part because their architecture was so much simpler. Unlike lasers, sensors were well understood. Since there were no big problems to solve, sensor technology was not considered to be a very glamorous area of research. Most development work focused on refinements in the materials used to prepare the devices, their electrical reliability, and the packaging needed to sustain operation in harsh environments.

One of the places where this work had been carried out was my organization at RCA Laboratories (now Sarnoff Corporation). Eventually two senior scientists who had been involved in developing the technology became interested in starting a sensor company.

This proved to be an attractive proposition for Warburg Pincus. The investment required to start manufacturing these components was low. Although it involved highly skilled, handcrafted work, there was no need for massive equipment of the kind used to produce integrated circuits. With the fundamental designs already in place, we believed there was a promising future for a new company that specialized in producing high volumes of light sensors for a variety of applications.

We knew that the sensors would have to be customized to each system manufacturer's requirements. But our thesis was that adapting the detectors to specific optical system applications would not require expensive modifications to the basic designs. Modest engineering changes around a standard architecture would get the job done. Moreover, the experience gained in fitting products to different applications would ultimately make the company a world leader in the field, with unmatched levels of expertise and cost efficiency.

In 1984 the two scientists, Drs. Greg Olsen (who became CEO) and Vladimir Ban (who became chief technology officer), left RCA Labs to found Epitaxx, the new company we had envisioned. They quickly identified the business management talent that needed to be recruited. This rapid addition of the right people served the company well in spurring its rapid growth and profitability.

The company acquired rights to RCA's detector technology. The strategy was to focus on a narrow product sector (detectors), maintain technological leadership, and produce devices profitably at the lowest cost, so system vendors would be willing to buy the components from a new company.

As the first business in the industry to focus exclusively on light sensors for optical communications, Epitaxx got off to a good start. Since there was no competition in the open market, at least initially, it could charge high prices for sample quantities sold to equipment manufacturers for system qualification.

Ultimately, however, its success would rest on its ability to supply superior products at a cost that could not be matched by the internal resources of the major equipment suppliers. The ability to embed customized components was also attractive to big system companies, as it allowed them to differentiate their products with unique features.

With the market growing rapidly as optical communications systems mushroomed, Epitaxx pursued a strategy of developing products and manufacturing techniques that continuously reduced costs while increasing quality. In Chapter 3 we noted the cost reductions to be expected in the electronics industry. Riding the cost curve downward proved critical, just as we had expected. Devices that sold for about $100 in 1984, when the market was starting to pick up speed, wound up being sold for only $10 just 3 years later. Because its manufacturing technology evolved to cope with rising volumes and dropping prices, the company was able to make a profit even at that low price.

The management of the company proved to be very resourceful in developing and marketing evolutionary technologies that kept Epitaxx ahead of competitors. For one thing, they took advantage of collaborative research done at universities. Furthermore, through its single-minded focus on a narrow technology and its access to a worldwide customer base, the company was able to leverage its resources very effectively.

Epitaxx became profitable very quickly, in only about one year, and remained so as it grew. Revenues reached $10 million in 1990, a modest figure but within our expectations. It became the leading worldwide supplier of optical detectors for fiber optic systems, making it an attractive acquisition target for a large company.

In 1991, when Epitaxx was acquired by Nippon Sheet Glass, a large Japanese company, Warburg Pincus exited the investment with a return of about three times its $1.7 million investment. The company continued to prosper. In 1999 the Japanese company sold Epitaxx to JDS Uniphase, then the leading manufacturer of optical components in the world, for $400 million.

STARTUPS SELLING SERVICES

Deregulation, the deployment of digital telecommunications infrastructures, and the creation of many new service offerings opened many opportunities for entrepreneurs eager to launch service businesses. At Warburg Pincus we identified four potential areas of interest.

- *Companies providing implementation services.* Expertise regarding the requirements for building and maintaining the new digital infrastructure was a scarce commodity in the industry. Hence it made sense to consider investing in startup companies, staffed by highly skilled specialists, which would provide infrastructure services to the industry on a contract basis.
- *New telecommunications service providers*, offering local voice and low-data-rate transport (such as modem access) over rented copper lines, were also attracting attention from investors. These newcomers were called CLECs, or competitive local exchange carriers, and would be in head-to-head competition with ILECs, or incumbent local exchange carriers, who owned the access lines. The RBOCs, successors to the old AT&T, were the largest ILECs.
- *Providers of broadband connectivity to consumers.* A technology emerged in the early 1990s that enabled the economical offering of data services in the megabit-per-second range over existing copper phone lines. This technology became known as digital subscriber line (DSL)

service. One interesting investment opportunity was to build a company that would offer this service nationally using rented ILEC copper lines.

• *Services managing local number portability.* Since the FCC had mandated local number portability, a new service that had never been offered before, new vendors were needed to provide technology and manage the communications infrastructure to enable it.

Of the four potential fields for investment, we concluded that new CLECs banking on local phone service faced the longest odds. As mentioned earlier, Warburg Pincus had successfully funded two companies in the *long-distance* telephony market before the 1996 Act deregulated local access, but competing in the local telephony market presented too many new challenges.

Startups would be offering a commodity service over lines rented from the incumbent RBOCs, which owned the physical access to the subscribers' telephones. This was not a promising competitive scenario and, besides, we saw far too many such startups for even the most optimistic market numbers. Inevitably, these would be competing with big incumbents, which have much lower marginal costs.

After due consideration we funded companies in the other three categories.

• IGS, founded in 1996, used proprietary software to provide infrastructure technology services to the telecommunications industry.
• Covad Communications, also founded in 1996, provided DSL access, then a new broadband service, to homes and businesses over rented copper telephone lines.
• NeuStar, founded in 1999, owned and managed the national infrastructure for local number portability, and sold services to carriers based on that infrastructure.

We will review the thinking that went into each investment, discuss its history, and consider why each had a different outcome.

IGS: Implementation services using new technology
This company focused on implementing new infrastructure technology into large telecommunications carriers.

In 1996, in the wake of the Telecommunications Act and the introduction of new network technologies, it was safe to assume that carriers were going to need help to cope with the new burdens placed on their copper line infrastructure.

As carriers deployed data services and opened their copper lines to others on a rental basis, a whole new set of requirements for infrastructure management came into play. Providers now needed to track such information as the distance between customers and central offices, the quality of subscriber lines, and the presence of impediments (such as load coils) on given lines that would preclude high-speed data service.

One factor in our investment decision was that IGS had built a track record in this field before we invested. Just as important, the CEO had developed solid business experience managing a small company. Started by entrepreneurs with angel funding, IGS had used its proprietary software and professional expertise to help a major RBOC to generate revenue-producing value from subscriber line information.

In effect, the IGS technology replaced the paper records that were traditionally used to manage the line inventory. Our evaluation of the technology supported the claims made for its quality. So did conversations with experts in the carriers that would be its prospective customers.

To make the company's value proposition as an investment even more compelling, Lucent Technologies had agreed to market IGS software and services through its sales channels. It even advanced the company several million dollars in advance of its making any sales.

With no need to prove that IGS could sell a useful solution, the only question was how quickly the company would be able to find customers and generate significant revenues to move it toward profitability. Its primary objective was to sign up the big carriers as customers. The technology was not interesting to small carriers with less complex infrastructures.

In fact, few sales materialized. While IGS did have contracts with two customers, Time Warner Cable and Pacific Bell, other carriers elected either to tap internal engineering resources or to contract with large, established vendors for the necessary software and expertise. It was remarkable how quickly the established information technology vendors were able to train and field a large cadre of new experts once the business opportunities became clear.

The company's small sales base never exceeded a few million dollars annually, and its need to continue making investments in marketing and product development kept it from ever becoming profitable. When its limited sales resources did manage to unearth a potential customer, the prospect would often insist on seeing the value of the offering before signing a contract. IGS wound up doing a lot of free engineering work up front, in a pre-sale mode. Larger firms could afford to do this; IGS couldn't.

The nature of the market militated against the success of such a small company addressing the needs of large carrier customers. In 1992 IGS was sold off in pieces, resulting in a loss for its investors.

Covad Communications: Faster on-ramp to the Internet
In contrast to IGS, Covad Communications successfully addressed a large consumer market rather than a limited number of large carriers. And its timing was excellent.

The Telecommunications Act of 1996 not only mandated that the incumbent telephone companies give competitors access to their copper phone lines, but it also established very favorable fixed rental rates.

One of the most promising of the markets that arose in response to this development was providing consumers with faster Internet connections. In the early 1990s PC owners who wanted to log onto the Internet and access messaging services had to dial into the system over their regular phone service. They faced agonizingly slow downloads and uploads, at maximum speeds of 56 kilobits per second. That's several times slower than what even a mobile phone offers today. And users had to pay for every call.

DSL technology, which was then emerging commercially, offered an answer to increasing consumer demand for faster Internet access. What made the technology so exciting was that it allowed widely installed twisted-pair copper telephone lines to carry up to 1 megabit of data per second, provided the customer was within a certain distance from a local telephone switch. There was no dial-up. It was always on.

For a flat monthly rate carriers could provide a consumer experience far superior to that offered by the available analog modems. Market surveys showed that many people would sign up to get faster access to the Internet if the price was right (under $100 per month). In fact, the companies offering video services over cable to the home were just beginning to consider offering data access services over their networks, but this required substantial upgrades.

DSL technology was already undergoing trials by several carriers, and the results were promising, as we found out from industry consultants. But several reasons gave the incumbent carriers pause before committing to a rollout. The first was the high cost (up to $1,000 per user) of the required customer premises equipment. Second, the RBOCs anticipated performance problems in using the old copper lines for high-speed data transmission. Finally, the carriers lacked staff skilled in managing the rollout of digital services to consumers. As a result, the RBOCs were in no rush to offer the service.

This situation created an opening for an entrepreneurial company to beat the big carriers (and cable companies) to market. That was the genesis of Covad Communications.

In 1997 Warburg Pincus was approached by a team of experienced managers with a business plan aimed at starting a DSL communications service company. Their idea was to rent copper access lines from the RBOCs and offer DSL service nationally, starting with the San Francisco Bay area. They believed that, given the slow rate of DSL commitment by the RBOCs, a new company focused on this market would establish an early-mover advantage and thus create a big market footprint before serious competition arose. Such a company could be very valuable.

The potential of building a major new company was intriguing but two important technological questions had to be answered first. The first involved the cost of the equipment needed to deploy the service; the second was related to the ability of the existing phone lines to carry data at the rates proposed.

Through our relationship with Level One Communications (discussed in Chapter 5), a company that made networking chips, we were able to get an expert analysis of the equipment technology. The conclusion was that chip costs would decline, and the price of the DSL customer premises equipment would drop rapidly from over $1,000 to well under $200. At this cost level it would be possible to offer consumers a very attractively priced service.

The second question was answered by tapping experts in infrastructure. From the technical data we collected, we concluded that a large share of the existing copper lines owned by the RBOCs would be usable for DSL service as long as the customers were located within a few thousand feet from the local telephone central office. This meant that a national service reaching a large part of the population was technically practical. The main issue facing the investment was the high cost of rolling out a service with a national footprint. An eventual price tag of $1 billion was not out of the question but the costs could be controlled by regional implementations.

Right from the start we recognized that cost of the service was going to be a critical factor in profitability, and reaching an acceptable price point required that the company achieve a satisfactory scale before competition became intense.

Our investment thesis was based on an acceptable balance of risks and advantages in their business plan.

- The management team was of a high caliber, though light on telecommunications service experience.
- We foresaw a rapidly growing demand for broadband Internet access at the anticipated offering price.
- Assuming that sufficient capital was available, a new company could establish itself as a national provider of DSL service. We knew the RBOCs and cable companies would eventually roll out their own services, but by that time Covad would have a huge head start, with a mature software platform to give it an edge in providing first-class service. The key was to grow to a sufficient scale for competitive costs to be achieved. Latecomers would then be playing catch-up, and would find it hard to replicate the service provided by the company if it executed as promised.
- Prodded by the FCC, the RBOCs would make sure the leasing of their lines went smoothly.
- Covad would have to develop and implement new software to manage its service, because available commercial software was unlikely to fit the company's needs. Covad's software would manage interconnections to the telecommunications infrastructure to permit its customers to access the Internet, plus implement customer billing. This involved a level of technological development risk.
- While Covad would initially rely on venture capital to get started, there was a significant possibility that the large amounts of capital needed to build the company could be raised from the public markets or large corporations, as had been the case in the deployment of both cable TV and cellular telephony services. In fact, we were all stunned when Bear Stearns approached us with a very attractive bond deal just as we closed on the investment. In any case, the service rollout would be gated by the availability of capital and such capital appeared to be available.
- The founding CEO had agreed that he would help in recruiting a successor with the right experience level if the company grew as rapidly as anticipated. In fact, this happened after our investment, as Robert Knowling, an experienced executive from the telephone industry, replaced the founding CEO, Charles McMinn.

Given our confidence in the technology and our conviction that the timing was right for both the demand for the service and the ability to finance the company, we decided to be the major funders of Covad Communications. We fully realized that the company was in a race to deploy a high-quality national service before anyone else.

There were two key elements necessary to succeed. First, the company had to build its own infrastructure. It needed software systems, sales and marketing teams, and people skilled in customer service. This meant assembling a large staff long before it could start generating meaningful revenues. As it turned out, company management did such an excellent job of building a service infrastructure that Covad was able to offer a high-quality service to customers right from the start. This infrastructure remained its most significant advantage over new competitors.

Second, because so much capital was needed for a national rollout, Covad would have to begin raising funds from the public markets and other sources as soon as possible. The company started by providing the infrastructure for the service in the California Bay Area. As it proved successful, the board decided to expand the service to other regions.

Covad might have been first in the market, but soon other venture-backed companies were leaping into the fray. The major telephone and cable companies also started to accelerate their own broadband offerings. Nevertheless, Covad stayed ahead of them in the early years because of its head start in building a solid business infrastructure and its sophisticated technology for managing the service requirements.

In part because of the company's outstanding execution of its business plan, our assumptions about the ability to finance the company outside of the venture capital sources proved correct, at least in the early years. Corporate investors participated before public financing took place.

Covad gained access to public markets as well, mostly because of the extraordinarily favorable environment for funding such companies in the late 1990s. Investors believed the sky was the limit for the value of new businesses in telecommunications, especially those rolling out leading-edge products or services to a huge customer base. This was also the time when the Internet caught the attention of the general public.

By December 1999 Covad had deployed a large DSL infrastructure in the US. It operated in 1,100 central telephone offices in 29 metropolitan areas, and its network "passed" 27 million homes, which is the industry term meaning it had access to that many potential customers.

Revenues grew rapidly from $26,000 in 1997 to $66 million in 1999, financed largely through equity and debt offerings. But Covad was not profitable because it had to continue investing ahead of customer revenues.

This was the time when public markets loved new communications companies. The company had an IPO on NASDAQ in January 1999 which raised $132 million.

When we exited our investment at the end of 1999, Covad Communication's NASDAQ company valuation was $5.5 billion. This was some eighty-five times its annual revenues that year, a reflection of the speculative fever of the dot-com/telecom bubble. Warburg Pincus received net proceeds of about $1 billion on an investment of $8 million.

The driver for this exit is an interesting story which we will return to when we discuss NeuStar.

After 2000, the company saw its valuation drop sharply, a victim of the end of the bubble and the crash in public values for telecommunications companies. It could no longer access the capital markets to fund its growth, but continued as a public company until 2007, when it was acquired by a private equity group.

NeuStar: Change your carrier, keep your number

NeuStar is a service company on the other end of the business from Covad: rather than consumers, its customers are the thousands of small and large telephony providers (including cable companies) in the US.

As mentioned earlier, the Telecommunications Act of 1996 forced the incumbent telecommunications service providers to rent access to their copper telephone lines to competitive providers. However, the FCC recognized that telephone subscribers were unlikely to switch if it meant changing their phone numbers. Therefore, it was mandated that local numbers be ported from one carrier to another.

Neither the incumbent suppliers (in most cases, the RBOCs) nor the new competition (the CLECs) had systems in place to automate the switchover. To make up for this deficiency, it was mandated that the infrastructure for enabling local number portability would become the responsibility of one or more independent companies.

Under the aegis of the FCC, a US telecommunications consortium was put together to review proposals for providing the service. At the end of the competitive bidding process and initial trials, Lockheed Martin emerged as the winner of a contract to design and manage the technology necessary for number portability. Its selection was based on its superior technological competence.

But in one of the contradictions of a changing market, Lockheed Martin eventually had to divest itself of the contract. The company was moving into other aspects of the communications industry which put it in conflict with its contractual requirements. The managers of the project (which was losing a lot of money) were encouraged to find outside financial support.

In 1998, the internal team that had done the original work approached us about leaving Lockheed Martin to found a new company with funding from Warburg Pincus. We undertook the task of assessing this opportunity, ultimately deciding to make the investment.

Our funding of what became NeuStar was predicated on the belief that an independent company was the logical choice to develop and run an infrastructure to enable local number portability. The business was already in existence as part of Lockheed

Martin, and had made significant progress in building the technology necessary to do the job. Our investment thesis included the following elements.

- Under the leadership of Jeffrey Ganek, the management team that would transfer from Lockheed Martin to NeuStar had the talent necessary to take the business forward.
- NeuStar had an exclusive contract for a certain number of years, allowing it to establish itself.
- The company's technology was developed to the point where a credible business plan could be formulated. Furthermore, pricing for NeuStar's services, based on how many phone numbers were ported, had been negotiated for the full term of the initial contract. This eliminated one major risk of starting a new company: uncertain pricing.
- We assumed that the volume of telephone numbers ported would rise as new telecommunications service providers began luring people away from the RBOCs by offering lower phone rates.
- Furthermore, we knew at the time of the spinout that there was a strong possibility that the FCC would mandate phone number portability for wireless, as well as wireline, phones. The "churn" of mobile phone subscribers switching carriers to get new phones or better deals would significantly increase porting volumes.
- Finally, our investment thesis recognized that a company like NeuStar would be in the enviable position of sitting at the center of intercarrier communications in the US. We believed it could eventually offer additional services that leveraged its costly infrastructure. There were lots of ideas, but at the time of the investment we did not know what those services might be.

But if the upside potential was good, the risks were significant.

- It would take large investments to build the necessary infrastructure.
- There was no guarantee that enough phone numbers would be ported to ensure profitable operation or a good return on investment.
- NeuStar could lose its contract exclusivity if its performance did not meet the industry's stringent requirements.
- While the industry had agreed upon the company's price for its portability service, profitability was not guaranteed. That would depend on the volume of numbers ported.

Acquiring this business required a lengthy FCC approval process. Part of the outcome was the creation of a whole series of "neutrality" requirements for the investors. As a result, we had to sell our position in Covad Communications. Covad had the status of a public carrier, and large investors in NeuStar could not be simultaneously large owners of public carriers. We were more than happy to exit our Covad ownership at its then-current valuation.

Fortunately, after we closed on the investment at the end of 1999, there were few surprises. We did find that the funding needed to bring the technology to the level of quality mandated by the contract was higher than anticipated.

On the other hand, the revenues of the company ended up exceeding our expectations. The annual volume of numbers being ported was rising rapidly. Much of this growth came from wireless number portability, which had been mandated in 2003 by the FCC, just as we had anticipated.

In addition, the company leveraged its customer relationships and infrastructure to diversify its services. As a result, NeuStar's revenues rose from $68 million in 2000 to $165 million in 2004, and it was profitable.

The company had an IPO on the New York Stock Exchange in June 2005. By mid-2006, when the company was valued at $2.4 billion, we decided to exit the investment. Warburg Pincus received about seventeen times its investment of $70 million.

LESSONS LEARNED

We have described the development and outcomes of five very different companies, launched in one of the most explosive investment environments in history. All entered their emerging markets early, and none played the role of a follower. Each started with a technical edge in its market and had a vision of how to build a valuable business.

In every case these companies succeeded in delivering the product or service they were created to provide. Yet two of them failed to establish themselves in their markets. Here are some important lessons from these investments that may apply in other situations.

New industry standards may be the basis of building a new company but caution is needed

The electronics industry has prospered by developing product standards that enable interoperability of products from different vendors. These standards vary in their complexity and industry impact, but very often new standards can create new winners and losers in the market, depending on how quickly customers and vendors adapt their products to meet them.

New standards that have the potential to create new markets are tempting opportunities for startups to jump ahead of slower competitors. The story of Licom illustrates the risk in such a strategy. Before committing resources to product development, there needs to be reasonable certainty that the standard will have a commercial future. Licom built its product around a promising new industry standard which, unfortunately, turned out to have a commercially insignificant future.

But a viable strategy for leveraging major new standards can build a great company. In Chapter 5 we will discuss Level One Communications, which took advantage of a new industry standard, Ethernet over copper phone lines, to build a highly successful business.

Defensible niche markets can be very attractive

Many startups in big emerging markets are too ambitious in their product plans. While the markets look to be wide open, soon enough competitors will emerge and make life difficult. I selected Epitaxx for discussion to illustrate the value of picking a defensible, if limited, product set from which to build a profitable business.

Epitaxx was successful precisely because its product was not ground breaking. It focused on building a dominant position in a product niche, where it could succeed by producing devices in high volumes, thus rapidly reducing their costs.

Under this scenario, the vertically integrated companies would not be able to justify the effort to manufacture these detectors themselves. Besides, these companies did not feel that sensors played much of a role in differentiating their products, as long as some degree of customization was available to fit their systems. For this, they needed a reliable source. Because of its early entry into the market, Epitaxx became that source. It quickly built a technological base and cost structure that made it difficult for later entrants to underprice its offerings and still make a profit.

Gaining access to large customers: A big hill to climb

For young companies launching game-changing new products or services into an established industry dominated by big companies, the biggest obstacle is entrenched buying patterns. Newcomers need to establish credibility with large customers that already have a roster of trusted suppliers, and are traditionally averse to risk.

Selling mission-critical products to large telecommunications service companies is particularly difficult. These companies tend to have lengthy decision-making processes, and they place a great deal of emphasis on the historical reliability and financial stability of their vendors. Both Licom and IGS faced these hurdles in launching their businesses.

Success in this situation takes a uniquely valuable product, a lot of persistence, good marketing partners, and time to gain a foothold. Faced with the difficulty of direct selling, most startups attempt to market their products through established industry vendors with strong relationships with the potential customer base. The long-term value of such relationships is likely to be unpredictable, as our experience with IGS proved. Lucent's position in the industry made it the best possible partner for IGS. Yet the partnership did not work as hoped.

In Chapter 7 we will discuss another startup, SynQuest, which like IGS relied on large companies for access to customers. This arrangement lasted until its best partner decided to sell a competing product of its own, forcing SynQuest to build its own, very costly direct sales force.

Licom, on the other hand, tried to sell to large customers directly. Its novel product did manage to get the attention of prospective customers, but for the wrong reason: they were more interested in learning about a new technology then entering the industry than in buying equipment from a startup.

Being first to market is great – staying in the game is the hard part

Being first to market is great. Each of our five companies enjoyed that advantage. But success breeds competition, and the lower the barriers to entry, the faster new competitive threats will emerge. In the case of NeuStar, of course, this situation did not pertain. The company had a unique industry contract to provide a number portability service, and the complexity of the infrastructure needed to do so was a high barrier for competitors. By developing new service products, it expanded its revenues and position in the industry.

Such favorable conditions are rare. More companies find themselves in the Covad situation – an early-mover advantage in offering a service or product on the one hand, but relatively low barriers to entry on the other, easily surmounted by competitors with enough resources. Then the strategy becomes gaining a market share so large that the company can dominate its market on the basis of parameters that mean something to customers – service quality and cost.

Epitaxx accomplished that objective in its niche market. The company built its competitive advantage on its declining cost of production and the ability to customize its products with minimum incremental cost. Therefore, Epitaxx could deliver products precisely fitted to customer needs more economically than its competitors, and still make a good profit.

Timing is (almost) everything – especially in accessing growth capital

Rapidly growing companies need capital to finance their expansion. Access to public markets through an IPO is historically the way to obtain capital at the lowest cost. NeuStar was able to take advantage of its unique market position to execute an IPO at a high valuation in a favorable capital market. It was highly profitable and growing rapidly at the same time. Covad was also able to enjoy a highly successful IPO, even though it was not profitable at the time, because interest in DSL services was high and competition was still manageable.

But markets are fickle. Communications infrastructure companies like Covad, which require continuing capital investments to fund their growth, are particularly vulnerable to the hazards of public markets. After 2000, just as Covad was achieving significant market penetration, conditions outside its control kept it from consolidating its gains. The financial market for telecommunications companies fell off a cliff, leaving Covad with a partial network footprint, high cost levels, and no way to raise more capital.

Taking advantage of opportunities to raise capital on favorable terms is vital. You never know when the window will close.

Nothing is forever – exits must be planned

Businesses must either periodically reinvent themselves or risk stagnation and decline. It is the responsibility of the board of directors to see that the assets of the company are preserved and enhanced.

In a world where technology migration is a way of life, and consolidation of industries around a few large companies is inevitable, the choices open to most boards come down to two: keep the company independent (and become an acquirer), or let it be acquired and maximize shareholder value in the short run. The first option requires access to capital, and the public markets are the preferred way to get that capital.

Epitaxx chose the second option: to be acquired. It thus avoided the need to expand its product footprint as the market evolved. By contrast, NeuStar has become an acquirer of smaller companies. Its business model and unique infrastructure make it easy to expand its service offerings.

Planning for the future of a business is all about objectives. In the majority of cases, entrepreneurs and investors aim at launching successful products or services within a time horizon that allows a profitable exit. This usually means selling the venture after a few years.

In rarer cases, the objective is to build a growing and profitable business over the long term – a business that will rank among the leaders in its sector. The strategy in this case is for the company to become a major force in its corner of the market through a process of internal development and acquisitions. Such companies can be thought of as "platform" businesses. There are limited examples of venture-funded businesses that achieve this status. NeuStar certainly fits that description.

In the world of venture capital, regardless of industry, companies with niche business models like Epitaxx can be successful if they can defend their position with leading-edge products. The capital requirements can be modest. They enter a market early, create value with pioneering products, and are later acquired by a larger firm. Investors receive a good return on their investment and the company contributes to human progress and economic growth.

If only it were as easy as it sounds.

5 Investing in a transformed market: Semiconductors

> It may be appropriate to speculate at this point about the future of transistor electronics ... It appears to most of the workers that an area has been opened up comparable to the entire area of vacuum and gas-discharge electronics ... It seems likely that many inventions unforeseen at present will be made ...[1]

You and I and every other person on earth are in debt to armies of tiny electronic servants that make the unparalleled quality of our modern life possible.

In 2009 each of us, no matter where we lived or who we were, had more than 10 billion transistors at our beck and call to help place phone calls, record and play video, reproduce music, conduct international business, even do astrophysical calculations. That is not a worldwide total – it is 10 billion per person. In 2003 there were – I hesitate to say "only" – 100 million per person.

That is astonishing growth, and the end is nowhere in sight. Today the transistor is the core processing unit in electronics. Practically every electronic product now in use is enabled by it.

It has evolved from a discrete device – normally packaged in a steel or plastic canister a little larger than a match head, equipped with three wires for connection into an electronic circuit – into an invisibly small component embedded in a self-contained integrated circuit (IC). Many millions of transistors can be etched onto a silicon (semiconductor) chip the size of a fingernail to create an IC with immense power.

So that's where those 10 billion transistors are hiding: in the IC chips in our mobile phones, PCs, GPS navigation systems, TVs, automotive engine controls, communication networks, and indeed

---

[1] W. Shockley, "An invited essay on transistor business," *Proceedings of the IRE*, 28 (June, 1958), 954. Dr. Shockley was the co-recipient of the Nobel Prize for the invention of the transistor.

all our electronics. The semiconductor industry keeps shrinking them smaller and smaller and packing more of them into ever more powerful ICs.

The economic value of the industry that created these remarkable devices is immense.

INVESTMENT POTENTIAL

From its inception the field of semiconductors has attracted a great deal of capital from private groups, corporations, and government agencies. I had grown up with this remarkable industry at RCA. When I joined Warburg Pincus in 1983 the idea of investing in semiconductor companies seemed obvious to me.

While I knew the technology and its potential from first-hand experience, I recognized that picking winning investments was going to be a daunting task. For one thing, the technology was advancing rapidly, and we had to stay on top of the latest developments. For another, massive investments in the industry, most of them overseas, were changing the competitive climate.

The semiconductor industry may have gotten its start in the US in the 1950s, but its strategic value as a national industry was quickly and widely recognized around the world. Starting in the 1960s and 1970s huge investments in Japan and Europe spawned big semiconductor competitors possessed of enormous financial and technical resources.

As venture capitalists, our chief challenge was navigating these troubled waters and producing attractive financial returns from our investments in this industry.

This chapter starts with a review of the industry, including our reasons for concentrating on the IC chip sector. With the stage thus set, we will review three representative investments that we made prior to and during a period of rapid growth in the industry. Then we conclude with what we learned. I believe these lessons will continue to be useful in evaluating this class of investments.

UNDERSTANDING THE INDUSTRY

To get an idea of the enormous advances that have taken place in the semiconductor industry over the years, look at how its basic materials have changed since 1947. Early transistors were all made of germanium, but today silicon is by far the most widely used semiconductor. Other, more exotic semiconductors are used for special applications where they offer advantages over silicon. To cite just one example, the chips that broadcast the wireless signals from mobile phones are generally based on gallium arsenide compounds.

Even more impressive is the transformation of size and form factor. Discrete transistors are still produced in a huge variety of sizes to suit various applications. However, the overwhelming number of transistors is found in ICs. The way they are interconnected in manufacturing the chip generally determines the functions they perform.

The price of these chips can range from tens of cents to hundreds of dollars, depending on their size and complexity. What is amazing is that chips selling for a few dollars can now perform operations that required mainframe computers in the 1970s. How these wonders were accomplished will be discussed below.

Products

You don't invest in an industry. You invest in specific companies. But you have to understand where their products fit within their industry. The semiconductor industry is broadly divided into seven major product groups.

- *Individual transistors* are used in power management and simple amplification or switching applications.
- *Microprocessors*, which are standardized 8-, 32-, or 64-bit chips, process digital data according to equipment-specific software programs. These have supplied the computing power for practically all electronic systems built since the 1980s.
- *Memory chips* store digital information. Volatile memories require power to retain the information. Non-volatile memories (commonly

called flash memories) have the unique ability to retain data after power has been removed from the chip.

- *Application-specific standard product (ASSP) chips* are dedicated to well-defined generic processing tasks. These include chips that process and amplify voice signals, convert analog to digital signals (or vice versa) to enable telephony or video, or enable digital data networking.
- *Programmable logic devices* are standard chips that can be electrically programmed by the customer to fit specific functional requirements. These devices, which allow equipment designers to very quickly build prototypes using proprietary logic, have greatly increased the rate at which new products are brought to market.
- *System-on-chip (SoC)* devices combine microprocessors and memories plus special circuits into a single, highly complex chip that saves space and reduces the cost of electronic systems.
- *Custom-designed chips* contain circuits for special processing requirements that are unavailable in commercial devices.

SoCs can be characterized as highly sophisticated application-specific devices. As electronic products continue to add features and functions, the processing power of SoCs has made them the fastest-growing product sector in the semiconductor industry. Custom chips, on the other hand, represent a sharply declining sector. They are being superseded by application-specific, programmable, and SoC chips, which offer equivalent performance without the expense of custom production runs.

Moore's Law: More processing, lower cost

None of these devices, nor the semiconductor industry itself, would loom large in the minds of investors were it not for the rapid establishment of the transistor and the IC as indispensable components of modern life. For this to happen, the industry had to transform the IC from an exotic specialty item into a viable mass-market commercial family of products.

As far back as 1965 one prominent industry figure saw how this could happen. That was Gordon Moore, co-founder of Intel

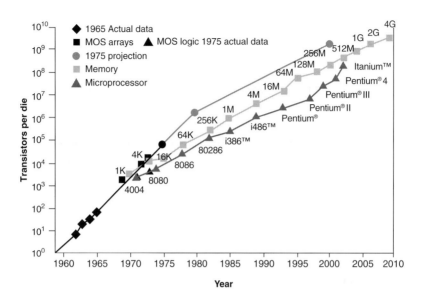

FIGURE 5.1. The increasing complexity of integrated circuit chips since the early 1960s. Shown here are the number of transistors per chip and the Intel microprocessors offered each year which take advantage of the evolving technology. *Source:* Intel.

(a venture capital-backed company), which grew into the world leader in microprocessor chips. He predicted that the number of transistors on a chip (i.e., transistor density) would double approximately every 18 months because the size of the individual transistors would continually shrink for the foreseeable future.

More transistors on a chip means higher processing power. It also allows the integration of functions that had hitherto required separate chips. Equally important, shrinking the transistors has the effect of making them switch faster, with a corresponding increase in data processing speed.

"Moore's Law" has held true for over 40 years, and the technical progress it describes shows no signs of faltering. Data from Intel in Figure 5.1 track the number of transistors on its chips from the early 1960s to 2008. The number has increased from barely 10 to nearly *4 billion*. The chart also shows the resulting growth in the processing power of Intel microprocessors.

Another way of understanding the impact of these advances in technology is to look at the decline in the price of each transistor in an IC chip (die) over time. In 1968 the average price was $1. It is now lower by a factor of nearly one *billion*. An individual transistor is essentially free.

Every product that uses ICs and microprocessors has benefited from this remarkable evolution in technology. With transistor counts per chip steadily rising, designers can put more and more functions onto a single SoC. As a result, consumers get a higher value product at a lower price.

The steady increase in chip performance, matched by a constant decrease in cost, has produced a unique win-win scenario for the semiconductor industry and its customers. The industry's market expands, while customers get ready access to ever more sophisticated, constantly more affordable electronics. This is the foundation of our digital world.

Cycles and growth

Any industry that grows the way semiconductors did is bound to attract the attention of investors. Figure 5.2 shows the industry's revenue growth, with projections to 2013.[2] Driven by the ever-expanding use of silicon devices in consumer and industrial products, growth has *averaged* about 10 percent per year – but you have to take into account the cyclicality of the industry, particularly after 2000.

This is a cyclical industry, whose fortunes depend on the ups and downs of industrial activity. Chips are practically everywhere, from heavy machinery to toys, and thus the semiconductor industry will track the business cycles of the larger economy. As the industry matures over time and semiconductor devices gain high penetration into the economy, it is natural that its overall growth rate, without

---

[2] B. Lewis, "Dataquest alert: Forecast, semiconductor revenue, worldwide, 2009–2013" (February 24, 2009), p. 3. Historical data provided through individual market share reports by Gartner, Inc. Forecast data provided by Gartner, Inc.

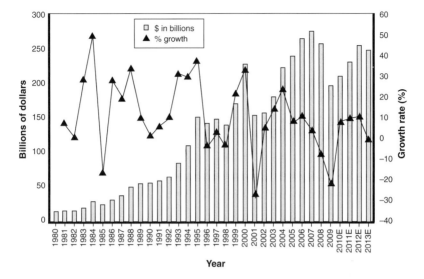

FIGURE 5.2. The global revenues of the semiconductor industry and their annual growth rate since 1980. The values after 2008 are projections. *Source:* Lewis (2009); data provided by Gartner, Inc. (ref. 2).

accounting for business cycles, will draw closer to that of the global industrial sector.

As is true with every industry, the growth numbers for semiconductors hide the fact that some product sectors have grown much faster than the average, while others have declined. The growth areas are where some startups have built viable business models, and where we made our investments. That helps explain why Level One Communications and Zilog, two of our investments discussed below, were able to grow at annual rates well in excess of 20 percent.

Opportunities for startups have come mostly from the shift from custom chips to application-specific ICs and SoC products. In the early days of the industry, equipment vendors commonly used custom-designed chips in their products to enable proprietary features. As the cost of designing and building custom chips has risen, these vendors have increasingly migrated to more standardized – and less expensive – SoCs designed for their product sector. These devices can be customized to their unique requirements through software programs. Our semiconductor investments were in these areas.

Global expansion, foreign rivals

Another factor that makes it hard to identify promising investments in the semiconductor industry is its rapid expansion overseas and the competition that has ensued. Though born in the US, the industry now counts many foreign-based companies among its largest and most successful enterprises.

Investments in the billions of dollars in Japan, South Korea, and Europe enabled the growth of giant semiconductor vendors such as Fujitsu and Toshiba in Japan, LG and Samsung Electronics in South Korea, Siemens (now Infineon) in Germany, ST Microelectronics in France and Italy, and Philips (now NXP) in the Netherlands. These companies supply memory chips and other products worldwide. Samsung in particular is a global leader in memory products.

Big as these giant corporations are, they, too, face increasing competition from relative newcomers. Starting in the 1990s, startups in Taiwan began to successfully penetrate the application-specific chip market. Since then some of these late arrivals have also prospered. MediaTek, for example, had 2008 revenues in excess of $1 billion and is listed on the Taiwan stock exchange.

When electronics manufacturing began moving to Asia, so did the market for semiconductors. The Asian market shift resulted not only from local manufacturers of branded products, such as Samsung, but also from companies that assembled products under contract for such US and European equipment vendors as Cisco Systems and Alcatel. Between 2001 and 2008 the portion of semiconductor output that went to equipment produced in the Asia-Pacific region increased from 25% to about 48%.

This shift in customer location has important implications for all semiconductor vendors. It means adopting a different approach to customer service.

As long as equipment makers were primarily located in the US, chip suppliers could focus exclusively on IC designs and leave the enabling software to their customers. US equipment manufacturers prefer to develop their own software for the application-specific chips that they use in their equipment.

By contrast, many Asia-Pacific manufacturers rely on their chip vendors for the software, and they expect to get it free. Software support is an increasing part of the cost of doing business for chip vendors serving this market. It is one more thing for investors to take into account in deciding whether to fund a chip company.

Symbiosis: Semiconductor startups and contract manufacturers

My personal experience in semiconductors taught me one important precept about the industry: the only constant was change. This was especially apparent in the rapid evolution of IC production techniques.

I began my career at RCA by designing one of the earliest mass-produced silicon transistors (the 2N2102) and putting it into production, all between 1960 and 1962. At that time there was no commercial equipment for semiconductor production, so we had to bootstrap our own. We not only designed the manufacturing process, we built the production equipment.

The same scenario played out on a much larger scale when ICs emerged later in the 1960s. There was no production equipment, so companies created proprietary manufacturing capabilities. They made a virtue of necessity by using their production skills as a competitive advantage. The cost of ramping up a factory also served as a barrier to entry for smaller rivals.

In the 1970s the situation changed drastically. Specialist equipment vendors such as Applied Materials began offering automated production systems, which grew more sophisticated by the year. Semiconductor companies wound up relying on outside vendors to supply their production equipment.

This meant semiconductor companies could standardize IC manufacturing technology around the machines they acquired. The benefits were significant. Without the need to commit internal resources to equipment development, they were able to bring plants on line faster, and concentrate on designing and producing ever more complex chips.

Standardization of production methodologies became a fact of life across the industry in the 1980s, when the CMOS (Complementary Metal Oxide Semiconductor) process emerged as the preferred technology for manufacturing digital chips. With the industry migrating toward CMOS, another group of specialist vendors emerged: chip contract manufacturers, better known as "foundries." The sole business of foundries is the production of chips for semiconductor company clients willing to live with standard chip design rules. This opened the way for "fabless" semiconductor companies.

By 2008 the well-funded foundries that emerged in the 1980s had become a force in the industry, accounting for about 20 percent of all production. The most important of these are Taiwan Semiconductor Manufacturing Company (TSMC, founded 1987), United Microelectronics Corporation (UMC, 1980, also in Taiwan), and Chartered Semiconductor Production Company (Chartered, 1987) in Singapore. Since 2000 some state-of-the-art contract manufacturing plants have also emerged in China.

Given the multibillion-dollar cost of new high-volume semiconductor plants, product companies can justify doing their own production only if their volumes are enormous, if they build products that require proprietary processes, or if their processes have special requirements and can use older equipment technologies.

Actually, quite a few semiconductor companies still profitably manufacture chips in small plants without the costly state-of-the-art equipment of the big industry leaders. But very few startups today take the path of owning factories.

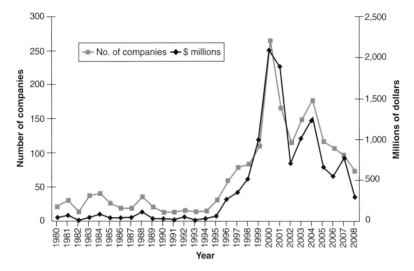

FIGURE 5.3. The number of semiconductor companies venture capital funded annually between 1980 and 2008 and the total venture capital they received annually. *Source:* Based on data from Thomson Reuters.

In short, for startups willing to work with standard processes, the playing field has been leveled. Contract manufacturers make it possible to launch a chip company even without its own IC fabrication plant. As Figure 5.3 shows, such "fabless" semiconductor firms began attracting capital investment in the mid-1990s. Investments in the sector peaked in 2000 with $2.1 billion divided among 264 such companies.

Technology meets economics

In this industry there is no place for new companies unless they are agile, creative, and fast – and well funded. The dominant theme is simple: integrate more functions into every chip you design so you can constantly add value to your customers' equipment while lowering their costs.

In the popular press this combination of higher value and lower price is regarded as the natural outcome of Moore's Law. But it does not happen without hard work. You need clever design to take advantage of the opportunities presented by the steady increase in IC processing capability, plus a deep understanding of customer equipment requirements. And of course your company must do all this profitably.

That is no easy task, as the history of the industry has shown. The semiconductor landscape is littered with the empty husks of chip businesses that failed to migrate their product lines as the requirements of their customers changed.

Nor are the economics of the industry getting easier. When chips had only a few thousand transistors, as was the case in the 1970s, their functions were very limited. It took many different kinds of chips to create a sophisticated system, producing high volume sales for IC makers.

Now, with many millions of transistors on a chip, makers have had to change their tactics. Their focus is on consolidating functions into one device, such as an SoC or application-specific IC, as a way to reduce the cost of the solutions they sell to their customers. They wind up selling fewer units, but at higher unit prices.

Meanwhile their customers, the equipment vendors, expand their markets and increase their manufacturing volumes as the functionality of their products improves. It is a happy synergy that yields profits for all parties.

But targeting the right market is critical for a startup. While some SoCs are relatively small and low in complexity, the cost to design a complex SoC chip with state-of-the-art technology can exceed $20 million. An investment of this size is only justified if the product generates large revenues – typically between $80 and $100 million. At the customer end, this translates to a relatively large volume of sales for the equipment that uses the SoC.

Paradoxically, it is just such barriers to entry that have spurred companies designing sophisticated SoCs and application-specific ICs to look for new ways to target their markets. First, as the number of interconnected transistors on a chip increases, system-level chip designers have developed more sophisticated skills to pack higher functionality onto these chips. In fact, chip companies have become the core contributors to system functionality.

Second, startups have had to become more creative with their products. As transistors shrink, the IC manufacturing process becomes more complex, and the cost of the production tooling increases. Tooling costs of a million dollars or more are typical for state-of-the-art products with the smallest feature sizes. At these investment levels startups can succeed only if they have unique products, good marketing teams, strong management, and strong financial backing. Those offering "me too" products generally fail.

There is no simple, surefire recipe for success. Only the most creative and best-managed companies can gain a long-term advantage. Successful companies learn how to build very efficient engineering teams, skilled in systems know-how, and maintain close relationships with their customers.

Being first with a novel and cost-effective chip is the best way of minimizing risk. Once a competitive chip is designed into a product family, displacing it is very difficult.

INVESTING IN CHIP COMPANIES

The investments we will discuss, all made before 2000, illustrate how three different companies faced the challenges of a rapidly growing industry.

We started the process of selecting investments with a decision to avoid companies in market sectors dominated by big companies with strong technology. We focused our attention on areas where entrants with limited capital could parlay novel intellectual property into successful products.

We decided to invest in companies that we thought would be capable of generating a sustainable, long-term product strategy. That way they could create enduring value as standalone companies, and would be candidates for public offerings in favorable markets.

To build sustainable businesses we looked for:

- experienced entrepreneurial teams able to manage a "platform" product strategy that they could expand to meet market needs and improve their technology;
- unique expertise and the creativity to generate a stream of market-leading products;
- a growing market not dominated by strong competitors;
- markets large enough to support a business with revenues well in excess of $100 million annually.

Our expectation was that such companies would grow to sufficient size and value to eventually become publicly traded.

It is not surprising that, of the hundreds of investment opportunities presented, very few fit our criteria. The great majority of deals proposed to us were for building a company on the basis of a single product, with no clear strategy for developing a larger product portfolio. In many cases the proposed product was likely to be late into a market where others had a head start. In other cases the entrepreneurial team was composed of engineers without the business and operational management experience needed to survive in this fiercely competitive industry.

We did find some companies that we believed to be consistent with our criteria. Each of those under discussion here addressed growing markets with superior technology and excellent entrepreneurial teams.

- Waferscale Integration offered chips incorporating an innovative non-volatile (flash) memory technology.
- Level One Communications produced application-specific chips that leveraged its technology for data networking over copper telephone lines.
- Zilog created SoC products for consumer electronics based on its proprietary microprocessors.

Yet only two of the companies achieved the success we envisioned for them.

WAFERSCALE INTEGRATION, INC. (WSI): FLASH
MEMORY-BASED CHIPS

WSI exemplifies how a promising startup with leading-edge technology products dependent on process innovations can be undermined by inadequate manufacturing capabilities.

As early as 1980 I became intrigued by EEPROMs (Electrically Erasable and Programmable Memories), then emerging as a technology for non-volatile memory chips. Such devices are now generically called "flash memories." Their unique value comes from their ability to retain information for years without the need for power. I actually started a program at RCA to develop a commercial product based on this technology.

It was not difficult to imagine the potential of these memories, but it took nearly two decades for them to realize their enormous promise. Flash memories are now a multibillion dollar industry. They are found in an enormous range of products, including digital cameras, USB thumb drives, MP3 players, and cell phones. As their cost has declined and their density has increased to gigabits per chip, they have started to replace small mechanical magnetic disk memories in personal computers.

The founders of WSI had invented a greatly improved non-volatile memory architecture. Their vision was to eventually design products that combined memory and data processing on a single chip, thus creating a subsystem that could connect to microprocessors and fit directly into data processing systems. Their first products were planned to be simpler memory chips. But this technology had not been reduced to practice.

The market potential for such devices promised to be in the hundreds of millions of dollars (which turned out to be an underestimate). But first, in order to realize these concepts in practice, the company had to solve some complex manufacturing technology problems.

WSI's entrepreneurial team consisted of talented technical and management people with experience at major semiconductor companies. A thorough examination of the proposed technology indicated that no fundamental roadblocks existed and that it was indeed novel and patentable. Therefore, together with several other venture capital firms, Warburg Pincus funded the company in 1984. Our contribution was $4 million.

Right from the start, everyone involved recognized that cost-effective manufacturing of these devices was going to be the key to a successful investment. Although the manufacturing technology for the chips shared many elements with the CMOS process then gaining ground in the industry, it required extensive modification and continuing innovation.

This would be possible only if WSI's engineers had overall control of the production process. So the question was, should the investors fund a new factory? That would have been the preferred course of action. Factories at that time were much less expensive than they are now, but it still would have required an investment in excess of $100 million plus continuing capital commitments to keep the factory state-of-the-art.

Had WSI started a few years later it might have been able to take advantage of the services of a contract foundry to manufacture the product. This would have required a special arrangement, because running special processes is not something that foundries like to do. Nor do they offer their customers' engineering staffs ready access to their facilities. In any case the foundry option did not exist in 1984.

So to avoid building its own plant, the WSI team proposed to strike a deal with a semiconductor company willing to accommodate "guest" products on its production lines. They convinced the Sharp Corporation in Japan to allow them to share the use of one of Sharp's production plants. In return they would license the technology to Sharp. The cost to WSI was reasonable, and all plans to build a factory were forgotten.

The company's first products were memories, but its real long-term focus was on chips called PSDs (Programmable System

Devices). PSDs were designed to add external, non-volatile data storage capacity to microprocessors from Intel, Motorola, and others. Although the product found customers, it never gained a large market, because it proved impossible to reduce its costs in line with industry pricing trends.

Cost proved to be WSI's Achilles heel. It was not possible to compete on memory devices alone when prices for ICs were dropping. PSDs were also affected by dropping prices, a subject we discussed in Chapter 3. While the industry was producing ever-denser ICs, WSI's products required the evolution of special process technology to allow the company to pack more components onto its chips. Implementing process changes in Japan proved to be difficult and time consuming. Given these constraints, it was not possible to aggressively pursue a strategy of increasing the sophistication of the chips while reducing their cost.

Eventually WSI found other manufacturing facilities, but the company was never able to generate sustainable revenue growth and profit margins. Several management changes occurred, but the basic issue of the company's lack of a manufacturing capability suited to its special requirements could not be solved.

WSI's annual revenues reached about $40 million at their peak, with no expectation that more growth was in the offing. In 2000, the company was acquired by ST Microelectronics, which valued its memory technology. We just recovered our initial investment.

LEVEL ONE COMMUNICATIONS: BUCKING THE FIBER OPTIC TREND

Level One Communications was a 1986 startup with the right technology at the right time. It also had a management team with the skills to take full advantage of the opportunity presented by the explosive growth in computer networking.

We were introduced to the company in 1987 by its newly appointed CEO, Dr. Robert Pepper. A brilliant technologist and businessman, he had been head of the RCA Solid State Division. In that capacity he had worked closely with me over a number of years to commercialize technologies that my teams had developed at RCA Laboratories.

When Dr. Pepper first described to us his idea for building a business around a novel technology for transmitting megabits of data per second over ordinary phone lines, we were highly skeptical. After all, I had spent years developing technologies for high-speed data transmission over fiber optics. The idea of investing in communications over phone lines sounded truly retrograde.

At that time, it was generally believed that multimegabit data traffic could be carried over long distances only over coaxial or fiber optic cables. Of course, homes and offices did not have such high-end wiring in place, and it would be very costly to add it. But every building was already wired with ordinary, unshielded twisted-pair copper lines for telephones. If these lines could also be used for high-speed data, that promised enormous advantages in business and home network deployment.

Unfortunately, signal attenuation over twisted-pair lines was high. This caused computer data error rates to reach unacceptable levels when transmitted through them for distances of hundreds of feet, as would normally be the case. Yet Dr. Pepper was sure that Level One could turn copper twisted-pair lines into a reliable medium for high-speed data.

He based his optimism on methods developed by the company's engineers for error correction on a chip, plus a superior technology for distinguishing transmitted signals from background electronic noise. By 1987 they had built mathematical models that simulated the effects of their technology, but had not yet built any functioning chips.

Convinced by his vision and an evaluation of the technology as it existed, we invested to fund Level One's first products. Although the investment was highly speculative, we had confidence in the leadership of Dr. Pepper and his engineers.

The risks were obvious. Would the technology actually work at the data rates and distances needed? Would the chip cost be acceptable? Would the company be early enough in the market to build a viable position before the inevitable competition appeared?

Most important, would the market be big enough? We could not gauge the size of this nascent market. It could be huge if data transmission over copper lines could greatly reduce or eliminate the need to install costly new coaxial cables. On the other hand, if the technology did not perform adequately and industry acceptance lagged, the market could prove too small to sustain a successful company.

Little did we dream that by 2002, copper lines would carry data rates as high as 10 gigabits per second over several hundred feet!

Unlike WSI's memory products, Level One's chips, which involved mixed analog and digital ICs, would *not* require proprietary production processes. The strategy was to evolve the products using standard manufacturing technologies available at contract foundries, which were then entering the market. This gave the company the flexibility of being able to use more than one factory in order to gain better leverage over production costs.

Since contract fabs were a relatively new phenomenon, we were concerned as to whether enough production capacity would be available when periods of peak industry demand caused a scarcity of manufacturing resources. In later years, when the company became profitable, it hedged against this risk. It made a minority investment in Chartered Semiconductor Manufacturing Company, a foundry in Singapore, to guarantee access to production capacity.

With manufacturing issues settled, the next big hurdle was waiting for the industry to settle on a standard for computer interconnections using copper lines. Level One had a unique skill set, but without established standards the market might be so fragmented among proprietary systems that profitability would be hard to achieve for any product line that Level One could market.

Choosing the standard around which to build the products was not a trivial undertaking. Two competing networking standards in the industry were receiving serous consideration, Token Ring and Ethernet. The company was reluctant to commit to creating connectivity products that would use copper lines while these two approaches battled for dominance. It wanted to be able to launch products that would address the interconnection requirements of the whole data networking industry.

While it waited for the winner to emerge from the standards war, Level One designed some custom connectivity chips for big equipment manufacturers. Its largest contract was with IBM, to produce a proprietary chip for their office network equipment, but IBM cancelled the project just as the chip was being completed.

After receiving $900,000 in termination charges from IBM in 1989, the company found itself losing money and without a single major product. In fact, Warburg Pincus had to commit additional funds to keep the company going.

We trusted in the basic technology thesis and the leadership of Dr. Pepper to find a successful path forward. But this refinancing was not a painless process. A number of early investors could not (or would not) participate, yet they blocked a new financing round we had proposed under terms that made sense to us. These matters were eventually settled by suitable negotiations and the company went on to great success. It proved to be worth the effort to refinance the company and wait for the right product opportunity.

In 1990 good fortune appeared in the form of IEEE 802.3i, a new standard for Ethernet data transmission over copper phone lines. Until then, the most widely used IEEE Ethernet standards had addressed only transmission over coaxial cable, and National Semiconductor Corporation dominated the market for chips that adhered to those older standards. Level One would soon replace that company as the vendor of choice in networking.

It is worth emphasizing how standards can cut both ways, and how vital it is to choose the winning one. As we saw in Chapter 4, backing an industry standard that was soon to be superseded had doomed Licom.

By contrast, after some earlier attempts to find acceptable Ethernet standards, the 802.3i standard was quickly accepted throughout the industry. This made Level One's success possible. Combined with the company's technological edge, the new standard allowed Level One to produce a winning connectivity product with a potentially huge market that included every interconnected computer in the world.

Once IEEE 802.3i was in place, the use of Ethernet transmission over coaxial cable faded away in favor of twisted-pair phone line networking. Demand for chips to handle the new scheme began to ramp up, and Level One took full advantage of the situation.

There were a number of reasons for its success. Most important, the company was early into the market with the best product available. Its superior technology, combined with the ability to deliver chips in high volumes at good prices, eventually won Level One about 40 percent of the world market for Ethernet 10 and 100 megabit-per-second connectivity chips.

The company also developed valuable intellectual property. Its strong patents were effective in preventing competitors from introducing copycat products.

The company matched its technological prowess with a savvy marketing strategy that helped it win and maintain a broadly diversified customer base, ranging from industry giants like Cisco Systems to network equipment startups in Taiwan. It also devoted a lot of attention to application engineering, which helped its customers quickly incorporate Level One products into their equipment.

Another company focus was leveraging its core technology to build a broad product portfolio. To supplement its products in enterprise computer networking, the company introduced chips for digital communications links between telephone central offices and businesses. This created a diversified customer base that included telecommunications equipment suppliers as well as computer manufacturers.

Level One's growth was largely driven by internally developed technology and products, but it did make two small acquisitions to acquire the engineering talent that it lacked in digital signal processing.

Financially speaking, the company was very successful. Revenues grew from $14.1 million in 1992, with a profit of $1.8 million, to $263 million in 1998 with profits of $36 million. Level One did its initial public offering (IPO) on NASDAQ in 1993, when its revenues were only $26 million. The company benefited from the fact that the public market investors loved fast-growing chip companies.

Warburg Pincus exited its investment in 1997 with a return of about eighteen times its investment of $11.5 million. Level One was acquired by Intel in 1999 for $2.2 billion.

ZILOG: SYSTEM-ON-CHIP FOR CONSUMER ELECTRONICS

Zilog was not a startup like the two companies discussed above, but I selected it to show how a failing business can be rebuilt by a first-class management team through a clever change in product strategy that leverages existing assets.

Our contrasting experiences with WSI and Level One Communications indicate that if a semiconductor company is to succeed, it needs to address large markets with innovative products that lead the industry. But it is best to avoid head-to-head competition with strong incumbents. And manufacturing capabilities must match the product strategy.

Our investment in Zilog proved that thesis.

How we got to invest in Zilog in 1989 is a story worth recounting. You never know when a chance encounter can generate a great investment.

Zilog was founded in 1974 by former employees of Intel. They were microprocessor innovators whose ambition was to compete with their former employer. Their first product was a proprietary

8-bit microprocessor, the Z80. An excellent performer for its time, the Z80 was used in early PCs, usually running the CP/M operating system that was popular before the advent of the IBM PC and MS-DOS.

Good as the Z80 was, competing head-to-head with Intel as a component supplier proved to be a bad idea. Intel's 8080 family of microprocessors came to dominate the 8-bit computing market. Having failed to build the company into a big-league component vendor, and in search of a very large target market, Zilog's management decided to enter the fiercely competitive computer business. This proved to be a really bad idea.

The fact that the business was doing poorly financially did not deter a rich buyer from becoming interested in acquiring Zilog. During the 1970s, the Exxon Corporation decided to expand from oil and gas production into electronics. It began acquiring companies that would become part of a new business unit named Exxon Enterprises. These companies were mostly on the equipment side of the industry.

Zilog, however, acquired in 1980, was a chip company. The fact that these acquisitions did not fit a coherent strategy became rapidly apparent as one company after another disappointed its owners.

Zilog was no exception. To revive Zilog, Exxon Enterprises hired Dr. Edgar Sack, a highly regarded industry veteran who put the company on the path to recovery. But by the late 1980s, Exxon decided to abandon its adventure in electronics altogether.

By the time Exxon started liquidating Exxon Enterprises, Zilog was profitable. Its annual revenues in 1989 were $90 million and it generated about $10 million of cash flow – a dramatic turnaround from the company's years of losses under its prior management. It was the last business left in the Exxon Enterprises portfolio, and Exxon management was in a hurry to sell.

We were introduced to Dr. Sack in late May 1989 by a mutual friend. Dr. Sack outlined his turn-around strategy of concentrating on proprietary Z80-based microcontroller chips for the consumer electronics market. These were basically small, low-cost computing "brains" that coupled the Z80 to other circuits on the chip to enable advanced functions in such products as television sets, TV remote controllers, wireless garage door openers, washing machine controllers, and cable set-top boxes.

Zilog's chips delivered high value at a low price. They sold for only a few dollars at most, far less than the more expensive and overly complex chips from other suppliers. This made them attractive for mass-market products. A washing machine doesn't need a high-powered processor to do a load of clothes.

We liked the strategy and, together with the management team, agreed to purchase Zilog from Exxon. We had come at the right time. Previously, Exxon had run an auction that failed to produce a transaction, so Zilog's management was able to choose a financial buyer, Warburg Pincus, as its partner. What's more, just as we entered negotiations, the relevant executives at Exxon – indeed, the entire company – were distracted by the disastrous *Valdez* tanker oil spill in the Gulf of Alaska.

The terms on which Warburg Pincus and management acquired Zilog were attractive: approximately four times free cash flow. Since we actually borrowed 60 percent of the purchase price from the Bank of Boston, our investment in Zilog counts as the first (very under) leveraged buy-out of a high-tech company.

These were the key assumptions underlying our investment thesis.

- The management team was well balanced, highly disciplined, and capable of carrying out a promising business strategy.
- The spinout from Exxon Enterprises was expected to be relatively simple in execution. The business was completely autonomous, with its own financial structures. Exxon provided no meaningful services.
- Consumer electronics design and manufacturing was moving to companies in Asia-Pacific countries, and those new entrants lacked the

extensive in-house engineering skills of established competitors. Thus they would rely heavily on chip vendors for the knowledge needed to implement their products. Zilog had a good head start in helping its customers design equipment around Zilog chips.

- Consumer electronics was a huge market, and the Asian opportunity would be pursued by much bigger companies. The expectation was that Zilog would remain competitive due to its excellence in manufacturing and customer support.

- The company's manufacturing plants were an important asset in pursuing a strategy that required flexible production at very low cost. These products did not require the most advanced production equipment, making plant ownership affordable. Zilog also owned efficient assembly plants in Asia. However, these plants had been neglected under Exxon management, so investment would be needed to modernize them. We were prepared to meet that requirement. What made such a commitment palatable was that we bought the company at a fraction of the depreciated cost of its factories.

The company executed well on its strategy.

Over the next few years, Zilog built a strong customer base among consumer electronics companies, particularly in the Asia-Pacific region. Samsung Electronics in South Korea and several large Chinese companies were early customers. Zilog's engineering, marketing, and sales organizations were trained to work closely with customers, helping them design products that incorporated Zilog chips.

Another source of strength was Zilog's ability to manufacture high volumes of chips at low cost, thanks to its flexible production facilities. Manufacturing excellence remained a top priority and the company continuously modernized its chip production plant in Boise, Idaho.

Zilog had its IPO in 1991 on the New York Stock Exchange. The company remained consistently profitable, although quarter-to-quarter variations related to the cyclical nature of its customers' needs affected its financial performance. Revenues reached $298 million in 1996, with a profit of $30 million.

In 1997 Zilog was acquired for about $520 million by a private equity firm and became a private company. Warburg Pincus received about twelve times its investment of $17 million.

LESSONS LEARNED

In our introductory overview of the semiconductor industry we discussed how certain historical patterns have impacted the fortunes of IC companies. Let's summarize the lessons learned from our experience, which may be useful to anyone considering investments in this industry.

Lock in the right production facilities

It has become fashionable to believe that the less a company owns of its production assets, the better. Why burden it with the capital costs of manufacturing plants if better alternatives exist?

Neither WSI nor Level One owned manufacturing plants, yet one did well and the other did not. Zilog did have its own plant, and was able to leverage this capability as a competitive asset.

To start with the least satisfactory case, WSI had talented entrepreneurs and valuable technology in a growing market. Yet after years of trying, it failed to build a viable business because its product strategy could not be executed without production facilities tuned to its unique requirements.

Because it did not control its own production, WSI could not meet the cost reduction demands that are standard in the industry. In spite of its novel technology, it was unable to compete in anything but a small market niche. It is noteworthy that one of the founders of WSI went on to found SanDisk, a very successful flash memory company that ranks among the world leaders in the field. But SanDisk had control of its production facility, in partnership with a large company in Japan.

On the flip side of the coin, Level One's technological edge came from its design skills, not from a proprietary production process. This meant its products could use the standard processes offered by state-of-the-art contract manufacturers. The ability to sell highly profitable products was based on chip functionality, plus the company's skill at incorporating functions on one chip that otherwise would have required several more costly components.

In the case of Zilog, owning its own factory was a competitive advantage. It gave the company the ability to be very flexible in serving a demanding set of customers. Consumer electronics is a business where orders need to be filled quickly. Building inventories is not a good way to meet this requirement because of the volatility in demand. Its flexible plant allowed Zilog to start and stop production as needed.

There is no single right answer regarding the manufacturing strategy. I highlight these issues because control of manufacturing plays a vital role in any business where the competitive advantage derives from process innovations or materials science. The same strictures apply, for example, to manufacturers of photovoltaic devices used in solar panels, and to makers of new materials, such as those based on nanotechnology.

The unstoppable move to increased functionality on a chip

As more of the value in electronic equipment moves into the silicon chips, the role of the semiconductor vendor has shifted from component provider to subsystem supplier. This new status means getting intimately involved in the design of its customers' products. In effect, the chip vendor becomes a "service" provider. Both Level One and Zilog owe much of their business success to their recognition of this new reality.

Zilog reinvented itself by meeting three conditions. First, it identified products that were valuable to its electronics equipment customers. Second, it acquired the engineering skills to design SoC products to meet customer needs. Third, its manufacturing costs were sufficiently low to meet the price expectations of its customers and still produce a profit.

Level One also used its strong applications engineering skills to get its products designed into new equipment, and to expand its customer base. Over time it built increasing levels of functionality into its products, based on the first-hand customer system knowledge its staff had acquired.

Pricing and product timing are as important as innovation

What happened to WSI and Level One demonstrates the validity of this old truism. WSI entered a market that had enormous potential for growth, as the success of flash memory devices has demonstrated during the past few years. Non-volatile chip memories are now a multibillion-dollar product sector in the semiconductor industry.

It had advanced technology and the advantage of being in the market early. Yet the company proved to be a poor investment because of its lack of satisfactory production facilities. This flaw kept it from delivering products in a timely manner at the right price.

Level One's history shows the benefit of having a true "first-mover" advantage: a proprietary technology that allows entry into a growing market defined by an accepted industry standard. The proliferation of computers led to a corresponding need for low-cost networking and ubiquitous connectivity. Copper phone lines were the cost-effective solution, if the technology could be made to work and the industry would agree on a standard. When Ethernet ended up the winner over proprietary protocols, Level One was there to launch outstanding products.

Competitors eventually matched Level One's original technology, but by then the company had built a solid reputation, an outstanding customer support organization, a broad product portfolio, and extensive customer relationships, all of which allowed it to deliver more value to its customers through incremental product innovations. This advantage would not have lasted long if Level One had not had a highly competent leader who could attract the operational talent to execute on new products while keeping quality and cost ahead of the competition.

Avoid frontal assaults on entrenched competition

There is another fruitful way to analyze why our three semiconductor investments met different destinies. We can examine how each one approached the problem of nullifying the advantages of the incumbent leaders in the markets they wanted to enter.

Level One displaced National Semiconductor as the prime Ethernet connectivity vendor because it recognized that a fundamental technology shift was about to take place. Ethernet over copper phone lines would soon replace systems based on coaxial cable, instantly allowing every PC near a phone jack to join a local-area network. Because National Semiconductor was wedded to coaxial, it lacked Level One's mastery of data transmission over copper twisted-pair. This is a textbook example of how startups build industry positions by taking advantage of major industry transitions.

Zilog's competitive edge was its early entry into an emerging market: Asian consumer electronics companies needed to tap its design expertise, not just for chips, but for the functions those chips enabled. Consumer electronic products were rapidly becoming more sophisticated, and the company was the first vendor in many customer sites just as they entered the market. Zilog offered state-of-the-art chip capabilities suited to the needs of its customers – at the right price.

As the Asian consumer electronics industry grew, big microprocessor vendors such as Motorola also recognized the opportunity it presented and entered the market with their products, but they did not have dedicated resources focused on the sector. Zilog was able to compete with them by carefully targeting its products to areas of competitive strength.

WSI, on the other hand, found itself battling in a huge market with the wrong weapons. It was in a mass-production, low-cost industry where it could compete only if its products used manufacturing resources to match those of its competitors. They didn't.

Gaining market entry gets harder as large customers
become scarcer in consolidating markets

The three companies we have been discussing were early entrants in their chosen markets in the 1980s and 1990s. Big changes have occurred since then that make the life of startups considerably more difficult. They center on the consolidation phenomenon which we discussed in Chapters 3 and 4. Consolidation is a way of life in this industry, and it greatly increases the barriers facing new entrants.

What makes the semiconductor industry so special is that its products are the controlling elements in practically every electronic system on the market. However, this enviable position carries with it some significant downsides for startups.

The biggest one is the higher barriers to entry. As chips become more complex, development costs increase, too. And as the industry consolidates, volume sales to underwrite the cost of product development are harder to find.

Investors in semiconductor companies thus face a double whammy. They have to commit to a much bigger investment to start a new semiconductor company than they would have in the boom days of the 1980s and 1990s. On top of that, the risk of failure is greater because too many suppliers are going after a shrinking base of large customers.

For example, personal computer (PC) and magnetic disk storage applications account for 40 percent of semiconductor industry revenues. But in 2008, there were only five major PC vendors. In the 1990s there were dozens. Similarly, there were dozens of disk drive manufacturers 15 years ago. In 2008 three equipment vendors shipped 75 percent of the volume.

The identical pattern can be seen in every electronics market segment. In the communications equipment arena Cisco Systems has emerged as the dominant worldwide supplier of networking equipment, with over 70 percent of the revenues. The merger of Lucent and Alcatel in 2006 brought together the two biggest industry players in telephone equipment. Two indigenous companies, Huawei Technologies and ZTE, have come to dominate the rapidly growing Chinese equipment market since the late 1990s.

In wireless handsets, another fast-growing industry sector, the five top vendors sold about 80% of all handsets produced in 2008. Nokia alone accounted for about 40% of these sales.

At this level of market consolidation, chip companies don't have many sales options. If they hope to grow in a significant way, they have to get at least one of the major players as a customer or focus on regional manufacturers in China.

That is not easy – the competition to "gain sockets" is intense, and keeping the sockets is just as difficult. New product cycles favor the vendor with a superior product over one with last year's technology, so winners and losers can easily exchange positions.

There is no doubt that the barriers to entry in the semiconductor field have become progressively higher over the past dozen years. This is typical of a maturing industry. It only highlights the need for selectivity in choosing an investment there. Investors must also anticipate the need for large capital commitments.

But the electronics industry is still very exciting technologically. It continues to create enormous value for its customers

6 Investing in early-stage technology: The Internet in the 1990s

> About the year 1450 some rather unusual 'manuscripts' made their appearance in the northern regions of Western Europe. Although not very different in appearance from traditional manuscripts, they were 'impressed' on paper, sometimes on vellum, with the mechanical aid of a printing press which used moveable type. The process was simple. But it was the object of considerable curiosity and fascination. In fact, these new books were to cause profound changes not only in the habits of thought but also in the working conditions of secular and religious scholars, the great readers of the time. The changes (we won't say revolution) soon broke the bounds of this original audience and made considerable impact on the world outside.[1]

Sometimes innovations trigger industrial and social revolutions. The introduction of printing opened up new vistas on the world, as printed books and periodicals brought knowledge and thought from every corner of the earth to those who could read. Without the printed word most of the great industrial advances and social movements of the past five centuries could not have taken root.

In our time the Internet has sparked a second revolution in the exchange of ideas. Its spread from a select few to the greater public reminds me of the way printing emerged: a slow start followed by spectacular impact. While the progress of the Internet has occurred in a highly compressed timeframe compared to that of printing (20 years instead of more than 200), they both spread from a specialist audience to the general public.

Before 1992 the use of the Internet was confined to the exchange of scholarly information among the scientific and academic communities. Although it had been operational for several years, its use

---

[1] L. Febvre and H-J. Martin, *The coming of the book: The impact of printing 1450–1800*, trans. D. Gerard (London: Verso, 1976), p. 9.

attracted little public attention. Few non-specialists were even aware of its existence.

Two milestone events opened the Internet to the general public. First, in 1992 e-mail service became available on the NSFNet Internet backbone. Second, in 1995 service providers like AOL, Prodigy, and CompuServe began offering commercial access.[2]

Of course, none of this would have mattered if fiber optic communications had not built up an abundance of digital bandwidth in the communications backbone (see Figure 4.1, p. 103). This gave the communications system the capacity it needed to carry Internet traffic just when the inception of commercial service opened the network to public data.

At this writing, some 13 years after the inauguration of commercial access, it is still hard to comprehend the extent to which the Internet has revolutionized human lives and expectations. For the first time in human history, people and businesses in far-flung locations can gather information and interact with each other instantaneously.

As a result, we can operate as a true global village on both personal and business levels. Social networking connects individuals, e-mail cements relationships, Websites promote myriad products and services worldwide, shoppers can buy from around the world, professionals can collaborate with colleagues on another continent. Miraculously, it all happens in real time.

We do not know what new direction the Internet will take in the future, or even how to maximize the social and economic benefits of its current capabilities. Imagine how chaotic the situation was back in 1995. Everyone was grasping for ideas, trying to understand how to harness an innovation that to all appearances was going to change the world. Our story begins there.

[2] W. Howe, "A brief history of the Internet," Walt Howe's Internet Learning Center, January 16, 2007; www.walthowe.com/navnet/history.html, accessed February 9, 2010.

INVESTING IN CYBERSPACE

It was not until 1995 that significant numbers of entrepreneurs emerged with new business ideas based on the Internet. A number of them found their way to Warburg Pincus in search of funding. By 1998 the early trickle had turned into a flood, and by 2000 into a tidal wave, as the Internet triggered one of the biggest investment bubbles in history.

We had a front-row seat for the emergence and rapid growth of the new medium. We were as impressed as everyone else with its potential. But we struggled to find investments that could be justified on the basis of their projected revenues and potential profitability. We were looking for startups that promised to become major businesses.

Subsequent events showed that our caution was well justified. As many people discovered, it proved to be considerably more difficult to build a profitable Internet business than the entrepreneurs and investors of those early days ever imagined.

Sorting the options

There turned out to be a lot of commonality among the proposals we reviewed.

As you might expect, all of the ideas presented to us were designed to exploit the Internet's fundamental capability: instantaneously and interactively reaching a very broad audience of users – essentially anyone with a computer and a phone line – and granting them access to practically unlimited amounts of stored information through widely available browser technology. Not only that, the proposals all clustered into five basic patterns:

- infrastructure technology providers that developed and sold the software and implementation services enterprises could use to incorporate the Internet into their business practices;
- connectivity service businesses that enabled Internet access from any computer;

- interconnection service providers that linked up companies over the Internet for specific functions, such as inventory tracking or financial settlements;
- online information publishers that operated Websites for users in specific social or professional groups. These startup ideas generally included browser capabilities for information searches;
- merchants that used the Internet exclusively for selling products or services, bypassing brick-and-mortar stores and traditional distribution channels.

There were many variants of these ideas. Online sales was especially popular, because practically any industrial sector could find ways of using the Internet to reach out directly to customers, thus avoiding the expenses associated with stores, direct mail promotions, tele-marketing, or sales calls.

All the entrepreneurs we talked with struggled to come up with profitable business models. They consistently underestimated the cost and difficulty of building an online customer base. Their proposed business plans, long on optimism but short on details, were built around one of three revenue models:

- direct payment from customers for products and services sold;
- subscription or transaction fees for services rendered or information provided;
- free services supported by advertising.

In essence, the entrepreneurs were saying, "trust us with some money and we will find a way to make a unique and very profitable business by leveraging the strength of the Internet before anyone else does it."

But while most of the proposals we saw were pretty shaky, it was not impossible to define and execute a viable Internet busi-ness strategy, even in the 1990s. We will begin by highlighting a few startups, classified according to the patterns above, to show how some early Internet businesses managed to build long-term value.

Infrastructure

Three Warbug Pincus investments that focused on infrastructure products and services are noteworthy for the leadership positions the companies ultimately attained. Their history is worth summarizing.

OpenVision/VERITAS: Warburg Pincus founded OpenVision Technologies in 1992 with a management team from Oracle to acquire, integrate, and market software utilities that made distributed, networked computer systems as reliable and secure as mainframe computers. Its key product made it possible to back up data across computer networks. OpenVision reached profitability and executed an initial public offering (IPO) in 1996 on $30 million of revenues. It then merged with VERITAS software, a similarly sized leader in software to manage data files within computers. The combined company became the unquestioned leader in data management software, with 1999 revenues in excess of $1 billion, just as the need for its offerings exploded with the growth of the Internet. In 1999 we realized $766 million on the $56 million Warburg Pincus investment, a thirteen times return on capital.

BEA Systems (a 1995 investment, largely divested in 2000–2001, with final shares sold in 2005) is now part of Oracle Corporation. It provides the software platform on which applications are built and run in distributed computing environments. BEA enjoyed phenomenal growth as the premier vendor of software platforms for high-volume electronic commerce over the Internet. It became a public company, traded on NASDAQ, in 1997, and in 2001 reached $1 billion in revenues more rapidly than any software company in history. The return on the Warburg Pincus investment of $54 million was $7 billion, a 130 times return on capital.

AsiaInfo (a 1997 investment, divested in 2006) grew into a leading system integrator of Internet and network technologies in China. The company became publicly traded on NASDAQ in 2003 with revenues of about $116 million. The return on the Warburg

Pincus investment of $12 million was $38 million, a 3.2 times return on capital.

Connectivity

We invested in only one company providing connectivity to the Internet: Covad Communications, a 1996 investment, divested in 1999. As discussed in Chapter 4, Covad became the early leader in delivering high-bandwidth digital subscriber line-based Internet access to consumers over standard copper telephone lines.

Interconnections

Warburg Pincus also invested in two companies that leveraged the Internet to enable proprietary business-to-business connections.

The Cobalt Group, a 1998 investment, sells technology and interconnection services that allow manufacturers and dealers to automate the automobile distribution process over the Internet. In 2008 it served about 12,000 dealers and 18 automobile manufacturers.

TradeCard, a 1999 investment, uses the Internet to automate the complex and expensive process by which letters of credit are issued to finance international trade.

Online information publishing

One business idea that generated quite a bit of excitement in the late 1990s was remote education over the Internet.

SkillSoft (a 1998 investment, divested in 2007) focused on enabling online employee management training. Here the Internet offers important advantages over conventional approaches. Instead of having to assemble in a classroom for training, enrollees can access courses and support from anywhere. This capability gives training greater reach into the organization. It also makes it possible for students to participate even while out of town on a business trip. The company became publicly traded on NASDAQ in 2000 and reached revenues of $217 million in 2006. On a Warburg

Pincus investment of $19.3 million, the return was $101.7 million, a 5.3 times return on capital.

Retail

We did not invest in dedicated online consumer retailers. We judged that the costs involved in building an e-commerce infrastructure were too high for a standalone Internet merchant. Online merchants would need to invest in marketing, brand building, stocking inventory, and shipping physical products. The high up-front costs and ongoing operating expenses would make it difficult to turn a profit.

In addition, we anticipated that existing catalog retailers or brick-and-mortar chains would adopt Internet commerce as fast as possible as a complement to their existing sales outlets. Since they were already generating cash flow from customers, and could avoid many up-front expenses by leveraging their existing infrastructure, they would have a huge advantage over an online-only startup.

Our concerns proved prophetic. The sole major Internet merchant success, Amazon.com, required huge capital investments and turned a profit only after years of losses. The most notorious of the online merchants, Pets.com, had to be liquidated just 9 months after its IPO when it could no longer raise money to cover its $300 million infrastructure investments.

Meanwhile Neiman Marcus, Target, Sears, and other brick-and-mortar retailing giants do lots of business on the Web.

CRITERIA FOR SELECTION

It is not surprising that early Internet companies attracted so many backers. When a new technology opens seemingly boundless horizons for hugely profitable markets, the first casualty among investors is rational thinking. They act on the hope that somehow circumstances will favor the most aggressive risk-takers. And to hear the entrepreneurs, every Internet proposal was a sure-fire financial hit.

It was obvious that some day soon the Internet was going to return enormous value. But how?

In the overheated environment of the early years, bubbling with ideas and extravagant forecasts of future success and with no historical context to calibrate the business risks, choosing promising investments required equal amounts of foresight and restraint.

We looked for startups that were clearly differentiated from the mass of proposals by a defined and credible revenue model. It was like looking for the needle in the proverbial haystack. While these were great objectives, it did not take long for us to discover that in the new Internet business world, there was no such thing as a well-defined business plan with known risks. In addition, so many ideas were floated, that even defining a sustainable competitive advantage was problematic. So we had to decide on the investment risks that we were prepared to take.

I have selected two companies for detailed discussion that illustrate business models uniquely built around the Internet. These enterprises also illustrate the value of what proved to be successful revenue models. EarthWeb migrated from a troubled advertising-supported site to a successful subscription model, while GlobalSpec relied primarily on subscription revenues right from the start.

EARTHWEB: INTERNET INFORMATION PUBLISHING

EarthWeb's history offers interesting insights into the development of a viable business model for the Internet. It is the story of how a 1995 startup focused on Internet publishing struggled toward profitability, going from an unsuccessful advertising-supported approach to a successful subscription model.

In 1995 advertising seemed like the most obvious way to generate revenue from Internet enterprises, for a couple of reasons.

First, it had worked for broadcasting. Radio and TV networks took in billions of dollars in advertising revenues to support their operations and, not incidentally, generate handsome profits. Why wouldn't this proven approach work just as well for online information providers?

Second, there was the matter of the ingrained culture of the Internet. People expected everything online to be free. Advertising offered a way to finance operations without confronting user resistance to paying for online services.

Initially the model delivered promising results. Even today the growth and the volume of online advertising continue to be impressive. In the US alone, online spending for 2008 was estimated at nearly $28 billion, and has grown for many years at a rate of over 20 percent per year. But Websites found that garnering a sufficient piece of that total is very difficult, especially since only $8 billion goes to non-search advertising.[3] As a result, advertising was never the sole basis for many successful businesses.

EarthWeb is a paradigm for the problems of trying to build an Internet business solely around advertising revenues. We will track it through several transitions in its business model and target market.

1995–1996: Consulting and Web hosting

EarthWeb was founded by three young entrepreneurs, Jack and Murray Hidary and Nova Spivak, who were captivated very early by the potential of the Internet to build new businesses. They believed that a business plan would have to evolve from experience rather than speculation. Hence, gaining experience from contract-funded work was as good a way to start as any.

In 1995 the three founders opened their doors as a Web development company, backed by money from angel investors. Their first priority, naturally enough, was to support themselves, and the only way they could do this was by contracting to provide Web services.

Their abilities and expertise won them Website development contracts from several large corporations, including U.S. West, a regional Bell operating company. Once they completed these projects, some of the customers asked the company to operate their Websites

---

[3] Statistics taken from K. Maddox, "IDC says U.S. online ad spending grew 20.1% in second quarter," www.btobonline.com/apps/pbcs.dll/article?AID=/20080825/FREE/991401 (accessed August 30, 2008).

as well. This service produced a few thousand dollars a month in recurring revenues.

While Web services provided enough revenue to keep the business going, barriers to entry into such a business were very low, and the founders knew that the field would soon be inundated with competitors. So they looked for a more interesting and sustainable business idea.

From early in its existence the company had focused its projects around the new Java programming language from Sun Microsystems. As a result, the technical team built up an impressive level of skill in the Java environment, which was quickly becoming a crucial part of Web development.

EarthWeb's founders sensed an opportunity to capitalize on these skills. Given the enormous interest in Java that had been building since 1995, they decided to start a Website devoted to information of interest to the rapidly growing Java user community.

Called Gamelan, the site drew a significant and steadily increasing number of visitors. They were attracted by free software applets (small programs designed to carry out specific useful tasks), as well as by the site's industry news and development support information. EarthWeb also established a relationship with Sun Microsystems, which was anxious to promote Java.

Gamelan enticed visitors by providing free information. The idea, replicated on a myriad of Websites then and now, was to increase the number of "eyeballs" on the site as an incentive for corporations to buy advertising.

Having built an audience (and uncovered potential customers), the founders also decided to offer a hosting service for "chat rooms," based on software that they had developed in the course of contract-funded work.

In its first full year of operation the company broke even. It made $500,000 in revenues and spent $500,000 on operations. While this was a good start, the management team was pondering a new direction, targeted at a much bigger and more profitable market.

The founders spent the first half of 1996 refining their plans. By June they felt that they had a promising business model, designed to appeal to a target customer base large enough to build a major enterprise.

As part of their strategic thinking, they also made a gutsy decision: they would stop soliciting contract-funded development work. They deemed this business too difficult to scale and too wide-open to competition. When they had completed their current contracts their only proven source of revenues would be gone.

Their new vision was to build EarthWeb into a dominant Internet-based marketing channel aimed at the rapidly growing community of corporate Web professionals. This target market was hungry for information, services, and products to help it develop Internet-based platforms for its employers. The company felt that it understood the needs of the potential customers in this target market.

EarthWeb would offer online "community activities," such as chat session hosting, technical content, and commercial software that users could buy right on the site. The revenue sources were expected to be a combination of advertising dollars and income from the software sales.

Because EarthWeb was a functioning company with thirty employees and a history, however short, of actually generating revenues, the founders were confident that they could attract the attention of a major venture capital firm. So they approached us with a business plan that called for annual revenues eventually to reach the hundreds of millions of dollars.

Such projections were, of course, highly speculative, but we decided to consider investing in EarthWeb for two reasons: the strength of the management team, and our judgment that the target customer base was not only growing rapidly, but had billions of dollars in purchasing power. The trick in turning EarthWeb into a successful company was to monetize its relationship with this community using then-unproven Internet access.

Our due diligence process showed that the EarthWeb team was very talented and that the software it had developed under contract had met customers' expectations in terms of costs and timing. On the business side, however, there was no way of projecting revenues and time frames any more realistic than those in the company's business plan. In cases like this, the strategy has to be to finance the company in stages and determine step by step when additional funding is warranted.

We talked to many people involved in Internet-related activities. All agreed that the Internet, a brand new interactive medium, had enormous commercial potential, but nobody would hazard an opinion as to how and when a profitable business could be built. (This did not stop many of them from offering us expensive consulting services.)

Even though EarthWeb's future was uncertain, it was clear that the Internet was going to be so valuable as a basis for building information businesses that a speculative investment was justified. The Internet represented a technological sea-change, and the audience identified by EarthWeb was a high-potential market for the right products and services offered via the new medium. The great unknowns were the magnitude of the advertising revenues EarthWeb could attract, and whether selling commercial software through a Website was going to be profitable.

1997: An advertising-supported Web portal

Having expanded its charter beyond Website services, EarthWeb became an information "portal" company serving the software developer community. Its new Website, developer.com, provided information on and listings of relevant products as well as "community" services such as chat sessions.

Developer.com's content and services were interesting enough to convince visitors to use the site. It offered technical information rather than industry news. Its goal was to attract software developers who needed rapid access to up-to-date technical data. EarthWeb's specialized chat sessions acted as a supplement to this information, allowing developers to form community relationships that would help them do their jobs.

Advertising provided developer.com's main source of revenue, but it continued to offer products for online purchase, including specialized software and books. Many of these items were not generally available through normal distribution channels, making the site a sole-source supplier. Rather than inventory and deliver products, the company relied on distributors to warehouse them and drop-ship orders to customers. Actual revenues from this commerce remained negligible, but the idea was to offer a convenient service to attract visitors to the site.

On the whole, providing content was the company's primary business, and advertising revenues its only meaningful source of income. EarthWeb spent a great deal of money filling developer.com with interesting and unique information. It hired domain experts to prepare the content hosted on the site, and also acquired other small Websites with content that it could use.

As a result, every month saw an increase in the number of page views. Most encouraging, with more eyeballs on the site, the number of advertisers began to grow, and the advertising rates it could command on the basis of Cost Per Million views (CPM) compared very favorably with that of other Websites.

1998: EarthWeb's initial public offering

In 1998, just 3 years from inception and 2 years after totally changing its business model, the company was consistently meeting its monthly revenue projections. Most of its advertising came from about a dozen large companies, including the likes of Microsoft, IBM, and Hewlett-Packard. EarthWeb had caught the eye of industry leaders.

It continued to acquire small technical Website companies, and signed content deals with a large number of information sources. These tactics insured that the quality of the site steadily increased. Page views were rising at a rate of 30 percent per month.

By June 1998, developer.com was the most popular site of its kind. Advertising revenues maintained their steady rise as the sales force matured and page views surged.

At this point the staff had grown from a low of thirty to a total of sixty. There was an excellent management team in finance, sales, operations, and content development. Although the company was still not profitable, its steady month-to-month revenue growth, coupled with increasing viewership, were grounds for optimism.

EarthWeb's progress peaked just as interest in Internet companies was heating up in the investment community. In August 1998 the founders began discussions with investment bankers regarding the possibility of a public stock offering for the company.

The stock was offered on NASDAQ on November 10, 1998 at an initial price of $14 per share. As part of the public offering, the company sold 2.2 million shares, yielding net proceeds of $27 million. Warburg Pincus and other early investors were also able to sell some of their shares. As a result, Warburg Pincus realized a small profit on its $17 million investment, while retaining additional ownership.

To get an idea of how irrational the public markets were at that time, consider the following. Within 2 days the stock hit a high of $86 per share, matching a 10-year-old mark for the biggest increase in the price of a new security. Two weeks later the stock price had settled in the $40–50 range, giving the company a valuation of between $300 and $400 million. The company also had two other public financings, for $26 million in May 1999 and $75 million in January 2000.

1999: Profitability proves elusive

The number of unique site visitors to developer.com grew rapidly, and EarthWeb's advertising revenues increased as a direct result. More than twenty million visitors logged onto the site in February 1999 alone. Revenues jumped from $3.3 million in 1998 to $31 million in 1999. Yet in that same year the company's cash loss rose from $9 million to $21 million, primarily because it was spending so much money to develop or acquire content.

Marketing expenses were also on the rise. A host of new Websites targeted at developers had emerged and were competing for traffic. EarthWeb had to beef up its content to hold onto its market share and maintain its attractiveness to advertisers. With the increasing competition from new Websites, and because it was not possible to measure the impact of the banner ads, the CPM price that advertisers were willing to pay started to drop.

2000: EarthWeb becomes Dice

EarthWeb might have wound up as another casualty of the collapse of the Internet bubble but for two fortunate circumstances: first, its content Websites were very popular; and, second, in the course of acquiring Websites related to its specialty, EarthWeb's management had bought one that stood out from all the rest. It was different, and it had performed amazingly well.

Dice, acquired in 1998, was a recruiting Website for information technology (IT) professionals. It had a much lower cost structure than the information publishing business. Most important, it operated on a subscription basis rather than being supported by advertising.

Dice grew rapidly under EarthWeb's management and became a leader in its niche. It was so successful that its sales figures soon surpassed those of the rest of the company. In 2000 Dice subscription revenues reached $46 million, while EarthWeb's advertising income dropped to only $17 million.

The board realized a change was needed. After it looked at all the options, the decision was made to exit the information business, give up the associated advertising revenues, and focus the company on Dice. No time was lost. The name EarthWeb, the developer. com Website, and all related assets were sold in December 2000. In January 2001, the company name was changed to Dice.

Scott Melland, an executive with extensive media industry experience, took over as CEO of the newly revamped company. One of his major tasks was to restructure the large debt burden the company had incurred during its $75 million financing in January 2000.

2001–2008: New business model

Online recruiting is a perfect example of how the Internet can fill an important market need better than existing solutions. In the Dice model, everything is done online for maximum efficiency. Dice receives resumés from engineers, screens them for accuracy, and enters them into a proprietary search engine. Its subscribers, consisting of both potential corporate employers and professional recruiters, use this resource to quickly find the talent that they need.

Subscribers benefit from timely access to a pre-screened pool of applicants. They can also measure the value of the Dice service by the number of interested applicants they find and hire. Unlike online advertising, Dice thus offers a clear indication of its return on their investment.

Dice's value to job-seeking engineers is just as compelling. It gives them a vehicle for making their availability known to a large number of companies at once. Because it builds on the power of the network effect, Dice quickly became the preferred Web listing site for information technologists.

Dice became profitable on a cash basis in 2001, with revenues of $50 million. The company continued to grow by entering other sectors of the recruiting market, such as financial services professionals. In 2008 revenues reached $155 million with a net income of $15.4 million.

EarthWeb left another valuable legacy. Its content Websites were acquired by Forrester Research and, after many years, continue to provide valuable information to the IT community. They are supported by a combination of search and banner ad revenues.

GLOBALSPEC: SEARCH APPLIED TO INDUSTRIAL
PRODUCTS

GlobalSpec's story is interesting in that it represents an Internet business model that proved viable almost from the start. In large part this is because the company provides a service that is uniquely the result of the Internet's functionality.

The founders of the company were engineers who had spent many years at General Electric in Schenectady, New York. They set out to solve a problem that they and many thousands of their fellow product design engineers faced every day: the difficulty of quickly finding sources for the components they needed to complete a project.

For over a hundred years the standard response to this problem was to look for products and vendors in the *Thomas Register of American Manufacturers*. Once you had identified companies who could supply what you needed, you called or faxed them to determine availability and price.

The *Thomas Register* was the bible of engineers. As I remember from my own engineering days, it was considered indispensable. Published as a multivolume set of large, green books, it was an encyclopedia of products organized by type and manufacturer. It occupied several feet of shelf space, but that was better than the alternative, which was poring through stacks or whole cabinets of catalogs put out by vendors and distributors. And it had a great business model: the manufacturers had to pay to be listed and the engineers who needed to find their products had to pay to buy the books.

In 1996 GlobalSpec's founders came up with a way to greatly simplify the component search process. Their approach exploited the advantages of electronic storage and the interactive nature of the Internet to provide benefits that no print publication could match.

- GlobalSpec would host the catalogs of the vendors of the targeted product sectors on a Website that was instantaneously available to users no matter where they were.

- The catalogs could provide much more information about each product than the *Thomas Register*'s listings.
- New products from the various vendors would be added as soon as they became available. By contrast, updated versions of the *Thomas Register* came out only once a year.
- GlobalSpec's key differentiator would be its search software. When an engineer made a search request, the system would find all vendors of a particular product in its database, without the cumbersome cross-indexing of the printed book.

Once the parameters of the service were established, the company's basic business model did not change. GlobalSpec has continued to extend its feature set and expand its product listings, but the original approach is still in place.

We decided to invest in the company in 1999 when it had just completed its first, angel-funded version of the software for catalog hosting and search. Today an engineer can search the site for a particular type of component, such as a resistor, and the "results" screen will show the various vendors selling the product. The engineer can request further information or quotes for delivery by clicking through to the vendor's Website right from GlobalSpec.

Registered engineers get free access to the stored catalogs. They also enjoy free access to periodic news updates announcing new products.

In 2009, with over 24,000 catalogs digitized and online and over 2 million components and 175 million searchable specifications in its database, the company is by far the largest such business in the world. It is a must-visit Website for information and products for engineers. All of this gives vendors a powerful incentive to have GlobalSpec host their catalogs. They pay a fee for the hosting service, but it buys them very broad exposure.

They also receive huge added value because GlobalSpec leverages the resources of the major search companies. When users search on any of over 600,000 keywords on Google, MSN, and Yahoo!, the search engines act as gateways to the GlobalSpec site, taking viewers right to the vendors' catalogs.

What's more, vendors can measure the value of the service by tracking the number of referrals that originate from the GlobalSpec site.

The company has become the dominant Internet business-to-business hub linking buyers and sellers of electrical, mechanical, and optical components. GlobalSpec estimated the 2008 market for components in its product coverage area at approximately $90 billion in the US alone.

With all these advantages it is not surprising that GlobalSpec displaced its print-based predecessor: the big, green *Thomas Register* books were discontinued in 2006. GlobalSpec's success was an early harbinger of a larger trend in information publishing. Nearly a decade after its launch we are seeing precipitous declines in circulation, advertising, and page count across all categories of print media.

LESSONS LEARNED

A gold rush with few big winners
The Internet is now an indispensable enabler of the world economy, but it has not made many investors happy. From our vantage point today it is easy to see why the Internet created such an entrepreneurial frenzy, and why the results were so disappointing.

There is no question that the Internet is an ideal platform on which to build services. As a medium of interactive communications, the basis of all commerce, it is unprecedented in its reach and power. In the early days it also had the lure of novelty, always a stimulant for a host of exciting, if unprofitable, new business ideas.

The thousands of entrepreneurs rushing to build Internet startups in the 1990s were not overly concerned about creating new, rock-solid business models. They gambled that they could reinvent practically any business by putting it on the Internet. Their mistake was in failing to consider whether migration to online operations made their businesses more competitive. Just because a service is on the Internet does not mean it has additional value to the consumer.

Investors and entrepreneurs also assumed that rapidly improving browser technology, which made it possible to access the burgeoning amount of information stored on the Web, would become an engine of commerce in and of itself. They ignored the fact that in the early days of the Internet this underlying technology had not been a commercial driver. It was used as a method to interchange scholarly reports.

In addition, everyone underestimated the difficulty of converting an established mindset. As mentioned earlier, making Internet users pay for what they were used to getting for free was a major obstacle, and there were few viable alternatives for generating revenue.

Finally, competition from existing "old economy" companies proved to be just as damaging to the dreams of entrepreneurs as that from other startups with similar business ideas. Entrepreneurs did not foresee how quickly and strongly existing retail businesses, for example, would move onto the Internet.

When the big, established merchants set up shop on the Web, newcomers found themselves facing entrenched incumbents with solid, sophisticated customer service infrastructures. Online retail merchants and providers of real estate listings, to name two of the most popular startup ideas, were especially vulnerable to this competition, and struggled to become profitable. Many of them never made it.

Finding the right Internet business model
What are some conclusions from the historical evolution of Internet companies?

New companies with enabling technology to sell were well paid for their products if they were truly leading edge. Several of the successful startups were providers of technology to run sophisticated Internet commercial operations. VERITAS, BEA Systems, and

AsiaInfo, for example, offered the technology and implementation services needed to rapidly build an Internet presence.

These companies had no problem getting paid for what they did, because they were delivering valuable, relatively unique products and services. You might think of them as modern descendants of the canny merchants who made money selling shovels to miners during the California Gold Rush.

Internet connections became a must-have service. Providers of such services, such as Covad Communications, or the companies that offered dial-up Internet access, also achieved initial success. However, they proved vulnerable to competition from the large telecommunications service providers, who considered Internet access an obvious extension of their communications backbone service. Once the giants woke up, new entrants found themselves fighting way above their weight class.

Companies that attracted a large customer base by offering a unique service that leveraged the interactive capabilities of the Internet were also able to prosper. GlobalSpec, Cobalt, TradeCard, Dice, and SkillSoft each eventually ranked among the leaders in their market niche.

The story was quite different for companies betting on the ad-supported model around which so many funding proposals and business plans continue to be built.

Why advertising-supported sites struggle

Ad-supported Websites were seen as the key to revenue generation in the 1990s. History has shown that profits from this approach are elusive, with very few exceptions. Yet entrepreneurs continue to be lured to advertising-supported business models even today. It is easy to understand why. With $28 billion spent on online advertising in 2008, and total ad dollars increasing at a rate in excess of 20 percent a year, it looks like there is more than enough money to go around.

That's an illusion. Nearly 75 percent of all Web advertising revenues go to Google, MSN, and Yahoo!, leaving thousands of other large and small sites to scramble for the rest. To understand why this is so, we have to distinguish among the different Web advertising models.

Websites in the early days of the Internet offered static display ads of various sizes and shapes. First there were banners. Then came skyscrapers (vertical ads), and ultimately pop-ups. When more people switched to high-speed connections that could handle animation and video, dynamic ads began to appear as well.

Advertisers were excited by the Web's ability to let them reach potential customers directly, on a one-to-one basis. They placed display ads on Websites that were in some way related to their product or service, and paid on the basis of the number of viewers who visited the site.

Underlying this payment model was the old concept of charging for ads by exposure. The value of an ad was assessed by the number of people, or "eyeballs," that would potentially read it. Advertisers were billed on a cost determined by the number of viewers (the CPM model). Theoretically, at least, this exposure builds "brand equity," familiarizing viewers with the company name and creating buyer preference for the advertised product. Hopefully it also drives some sales.

Print and broadcast outlets had used the same model for decades. The higher the circulation of a magazine, or the greater the ratings for a TV show, the more advertisers paid for placing an ad, because greater exposure presumably translated into more brand equity and hopefully more sales.

But number of viewers is a passive metric. It can be correlated to sales only indirectly, assuming there is any discernible effect at all. Advertisers found it impossible to measure the actual return they were getting on their investment in "image" or "brand equity" display ads.

In broadcast or print, which generally do not generate direct response from their audiences, that may be the best you can expect.[4] But in an interactive environment like the Internet this situation was intolerable, especially since a better solution was already available.

Click-through technology had been around since 1994. It not only allowed viewers of an ad to go directly to the Website for the product or service being promoted, it also gave advertisers a way to count the number of responses they were getting for their advertising dollar. Yet most sites continued to charge on the basis of CPM right up through the Internet advertising crash of 2000. It's no wonder that so many Internet companies found themselves facing an advertiser revolt and a revenue crisis.

The fragility of advertising-supported media should not have come as a surprise. In the media world, building businesses on the basis of advertising has always been difficult – witness the historical failure rates of newspapers and magazines. Internet entrepreneurs failed to learn from history. Worse, they repeated the mistakes of the past by not adapting their methods to the capabilities of the medium they were exploiting.

Obviously a new model was needed, and the click-through ad provided its technological underpinnings. Websites began to introduce advertising rate schemes that were more closely related to results. The two most prevalent "active" models are Cost per Click (CPC) and Cost per Acquisition or Action (CPA).

CPC charges advertisers on the basis of how many viewers use an ad's click-through function to visit the advertiser's Website. CPA measures how many users complete an action at the advertiser's site, such as buying something or signing up for a service.

---

[4] Direct response TV pitches ("Call now to order" spiels) and "infomercials" are an exception, but one that proves the rule. They are not the predominant mode of broadcast advertising. In print, reader response cards come closest to a direct response mechanism. They generate leads, but not sales.

While these new metrics helped advertisers determine true return on investment for their ads, they did not address two other shortcomings of Web advertising. First, the number of active Websites on the Internet was growing exponentially, reaching approximately 8 million in 2000 and standing at 71 million in August 2008.[5] This created a glut of available advertising space, and a lot of uncertainty as to where the best place for an ad really was. Rates dropped drastically in response.

Second, there was the difficulty of making sure the right visitors saw the ads at the right time – when they were actively looking for the advertiser's product or service. Since visitors came to informational Websites for various reasons, sites like EarthWeb could not guarantee that.

But search engines could. People went to a search engine precisely because they were looking for the item specified in their keywords. This fundamental characteristic gave search sites a huge advantage. It is the reason why search advertising monopolizes the majority of Web advertising dollars.

Google was the pioneer in monetizing the online search capability with new technology and sales models. Google advertisers contract to pay for one or more search keywords. When a Google user types one of those keywords into the search engine, it brings up an ad for the advertisers who "own" that keyword, along with the normal non-advertising hits.

The ads are placed prominently on the screen in an order predetermined by how much each advertiser has promised to pay. What they pay is determined by a very sophisticated auction architecture that actually represents Google's original "special sauce" (Google now augments this with a suite of proprietary technologies for deploying and managing the largest data centers in the commercial world). Advertisers only pay Google when someone actually clicks

---

[5] Statistics drawn from the Netcraft website (http://news.netcraft.com/archives/web_server_survey.html), accessed September 9, 2008.

through to their Websites. This system gives advertisers a direct measure of the impact of their ads.

GlobalSpec's success demonstrates the effectiveness of search advertising. It, too, is essentially a search site and a source of information to its subscribers. Its model works because there is a firm connection between site visitors and the originating listing or ad. Catalog owners can measure the value of their use of the GlobalSpec site in terms of sales leads generated.

That's a capability that is hard for non-search sites to match. However, it should be noted that banner ads are actually now contributing to GlobalSpec's revenues. The ads are featured in the e-mails that the company sends out free to its subscribers to describe new products.

Building stable revenues

We ended the story of EarthWeb with its transition to Dice, a subscription-supported engineering recruitment Website. Recruiting is a perfect example of how a business function can be transformed by the interactive capabilities of the Internet.

In the traditional recruitment process, employers place classified advertisements to attract job-seekers, then screen the resulting applications themselves. Or they use employment agencies and executive search firms as intermediaries to identify candidates for jobs. Both approaches get them involved in time-consuming back-and-forth communications. With the Internet, on the other hand, much of that wasted motion goes away. That is the value proposition for an online job site.

Since Dice is a major player in a niche job market, it attracts the best candidates and the most important potential employers. This classic network effect produces a sustainable competitive advantage. Applicants can list their resumés on Dice for free, but employers are willing to pay handsomely for the service because it produces better results, with less time and effort, than other means of acquiring candidate data.

If the company had continued as an information publisher and relied on banner advertising for its revenue, it probably would have failed, like so many other similar businesses.

True, things have changed since 1999. Display ads such as banners and pop-ups now offer interactive click-through features, too. It is worth considering how well EarthWeb's ad-supported model would have fared if these ads had been widespread at that time.

Click-through ads are an intriguing hybrid form unique to the Web. They meld the branding of a passive display ad to the measurable lead generation and sales potential of a direct response ad.

Chances are they would have delivered higher revenues for EarthWeb had it stayed with its original content business. However, given the continued struggles of ad-supported Websites in 2009, I doubt it would have made a difference in the long run.

CONCLUDING COMMENTS

At the beginning of this discussion I drew parallels between the invention of printing and the advent of the Internet. Both phenomena vastly expanded the scope of human understanding and the spread of commerce. Both created a faster, more efficient way of communicating among the masses of people.

Of course the analogy is not perfect. The Internet does not work the way print does. Print is a monologue, while the Internet is a dialogue. Print is basically one-way communication, and the Internet is fundamentally interactive.

There is another difference which, while it may not be fundamental to the respective natures of the two media, speaks volumes about the effect of the Internet on our economic and social response. Simply put, the Internet has developed with frightening speed, while print gave us time to get used to it.

By most estimates literacy took nearly 300 years to become widespread in Europe. (Experts arrive at this interval of time by tracking the dissemination of printing technology.) This does not mean literacy was universal in the eighteenth century, or even the nineteenth. In the mid-1800s over a third of the citizens of England were not literate enough to write their names on marriage licenses.

Contrast that slow percolation of the printed word with the wildfire spread of the Internet. In 1995, when the Internet was opened to commercial use, there were 16 million users. In the second quarter of 2008, nearly 1.5 *billion* people used the Internet.[6]

With a total world population of 6.5 billion, that means more than one out of five (22 percent) humans on the planet is on the Internet. In just 12 years the Internet has achieved the approximate level of penetration that print took three or four centuries to attain. And the comparisons don't stop there.

Today the admittedly less-than-definitive but highly indicative statistics collected from the nations of the world by UNESCO show literacy rates as high as 99% in central and eastern Europe, and as low as 60% in south Asia and sub-Saharan Africa. UNESCO estimates average overall world literacy at 79%.[7]

Internet usage could match that figure in another 10 years just by matching its growth of some 306% between 2000 and 2008. It has already assumed a central place in the lives of people in every part of the world. There are 51 million users in Africa, 42 million in the Middle East, 20 million in Oceania and Australia.

During the Internet's early days entrepreneurs, investors, and users were like the man riding on the back of the tiger. They couldn't control the beast, and didn't dare get off. The best they could do was react. Investors in particular felt they had to make their choices quickly, before the opportunities were gone.

---

[6] Internet statistics cited in this section are taken from www.internetworldstats. com, accessed September 2, 2008.

[7] UNESCO Institute for Statistics, 2007 figures.

That chaotic environment promoted the growth of unrealistic expectations on the part of both entrepreneurs and investors. The crash of 2000 simply highlighted how unrealistic everyone had been. But the lure of the Internet continues, and with it the challenges of finding stable revenue sources.[8] It is not getting easier.

[8] L. Gomes, "The pied piper of pay: Web entrepreneurs these days are getting the wrong idea that all that matters are users – not revenues," *Forbes* (June 22, 2009), 44.

7 Software products and services

Computer technology … is the technology of how to apply knowledge to action to achieve goals. That is, it provides the capability for intelligent behavior. That is why we process data with computers – to get answers to solve our problems … That is what algorithms and programs are all about – frozen action to be thawed when needed.[1]

Thirty years ago, before the dawn of the PC age, software was already a big business. Most of it was written for mainframes and mini-computers. Software vendors, including the computer companies, contracted with the corporations that owned this "big iron" to create or customize programs for their business needs. The general public was blissfully unaware of what software was and how it worked, and few understood the difference between software and hardware.

Now that most people use PCs, they know what software is and does. They may use a word processing or spreadsheet program at work, a genealogy or personal finance package at home, and an e-mail program and Web browser everywhere. There is a software program to enhance every aspect of modern life.

Packaged or "shrink-wrapped" programs such as games, word processors, or video editing suites are the most important products in today's consumer software market. But there is a big market for business applications, most of which are customized adaptations of commercial software products. Banks, airlines, government institutions, customer service organizations – they all rely on soft-ware to run their business processes, record orders, track activities, and more.

Given the enormous value created by innovative software, it is not surprising that software companies have been investor favorites for many years. Between 1980 and 2007 venture capital funds invested a total of about $94 billion in software companies, and they

---

[1] A. Newell, "Fairy tales," *Computer science: Reflections on the field/reflections from the field* (Washington, DC: National Academies Press, 2004), p. 184.

continue to be interested in the industry. As Figure 2.2 (p. 58) shows, 768 software companies received 17 percent (about $5 billion) of US venture capital commitments in 2008 alone.

Software companies are considered attractive investments because of their potential to generate revenue without incurring large capital expenditures. Software consists of computer code, a form of intellectual property, written by highly skilled and knowledge-able programming experts to accomplish specific tasks. The over-head required to create software products is minimal. Programmers need no production equipment other than computers, which are now commodity products. While they often work in teams, they do not necessarily need adjacent offices. Modern communications make it possible for them to work remotely.

Software manufacturing expenses are also low. The cost of packaging and distributing the product is being eliminated as more people choose to purchase software online and download both the program and the guide directly onto their computers rather than buy discs and books.

In the software business, scale matters. Investors frequently overlook the increasing cost of building profitable software compan-ies. Both marketing expenses and product development costs are ris-ing, driven by rapid technology changes and intense competition as the industry consolidates. As we saw in Chapter 3, this is a normal course of events.

Over the years Warburg Pincus has made investments in com-panies that based their business on software of various kinds. These included companies that developed and sold software as a product or implementation service, as well as service companies that dif-ferentiated themselves from the competition through proprietary software.

In the previous chapter I referred to three software startups in which we made investments. Two of them, OpenVision and BEA Systems, became leading providers of software that enabled Internet-based electronic commerce. The other investment, AsiaInfo, a Chinese startup, provided information technology implementation services.

Another noteworthy Warburg Pincus investment was Ness Technologies, an information technology implementation services provider. The company was started in 1999 with the merger of six Israeli software services companies. We later added two other Warburg Pincus portfolio companies to the mix, one in India and the other in the Czech Republic.

With revenues of $665 million in 2008, Ness, with its stock traded on NASDAQ, is the largest information technology services company in Israel. It also operates in sixteen countries in the Americas, Europe, and Asia.

I have chosen three other startup investments for discussion in this chapter. All were mentioned in connection with the assessment of technology risks in Chapter 3, but their cases merit a more detailed consideration. The diversity of their history allows us to draw useful lessons with broad applicability to the software business.

The first is Nova Corporation, a startup that built a highly successful credit card processing business on the basis of an innovative software and communications architecture to enable its services. The other two companies, Maxis and SynQuest, developed and sold software products targeting the entertainment and large enterprise markets, respectively.

NOVA CORPORATION: CREDIT CARD PROCESSING FOR SMALL MERCHANTS

Every entrepreneur at one time or other gets the same piece of sage advice: don't try to sell a commodity service in an industry that is dominated by well-funded giants.

Don't do it – that is, unless you have a business proposition so compelling that it overcomes the normal barriers to entry. Nova Corporation is an example of how such a venture can succeed.

In 1991 a group of managers with experience in the credit card industry decided to form a new company. Its competitive advantage would come from new technology that they believed would reduce processing costs and improve service. Since banks dominated the credit card market, the risk of backing the new company was larger than usual, but this one was targeting small merchants, a sector that was then poorly served. The team brought a number of advantages to the table: fresh ideas on technology, deep knowledge of the industry, and useful relationships that could leverage the market entry of a startup.

Such a venture faced obvious challenges from well-positioned incumbents. The team approached Warburg Pincus because they wanted a financial partner with the resources and patience to stay the course. We assumed that role, made the investment, and were ultimately well rewarded.

Nova is an example of how a successful service company can be built, even in a consolidating market, by combining outstanding technology and smart marketing. To understand how this was accomplished requires a brief introduction to the industry.

Background: How credit card processing works

Widespread use of third-party credit cards, as opposed to store charge cards, is a fairly recent phenomenon. Diners Club, founded in 1950, is often cited as the first modern card. However, the industry did not really begin to grow until 1958, when the American Express card and Bank of America's BankAmericard (later to become Visa) were introduced.

Initially these issuers maintained a tight control over the processing of the charges on their cards. But in 1966 Bank of America began franchising its Visa card to other banks. At just about the same time a competing group of banks introduced the MasterCharge (now MasterCard). Opening the system moved the service into the mass consumer market, and gave impetus to the expansion of credit card use.

Every bank within the Visa and MasterCard associations could issue its own co-branded version of the plastic cards to its customers.

Banks also signed up stores to accept the cards. Yet the cards were and are universal. Visa customers, for example, can shop at any store that accepts Visa cards, no matter which bank issued the card or holds the store's account. When the cardholder and the store deal with different banks, as is usually the case, processing the transaction requires transfers of funds between the banks.

With the sudden increase in the number of banks issuing cards after 1966, transaction processing steadily grew more complicated. Numerous parties were involved in clearing a transaction. The subsequent growth in consumer use of the cards added volume to complexity.

Banks began looking for ways to reduce internal technology investments. Some of them responded by handing off the processing function to outside vendors. This development opened the way for Nova's eventual entry into the market.

In the early 1990s, when we invested in Nova, credit card processing volume was growing at an annual rate of about 18%. Since 2000 volume increases have ranged between 8% and 10% per year. Today over $1 trillion of credit card charges are processed annually in the US.

It takes a robust and sophisticated data processing network to handle a load like this. Every time a customer's card is swiped through a store terminal, the network, along with its associated software, must close the loop from the initial transaction through the settlement of the account. This includes arranging payment to the merchant by the merchant's bank, transferring funds from the bank that issued the credit card to the merchant's bank, and issuing a statement to the customer.

Figure 7.1 shows a highly simplified schematic of the transaction process, indicating the steps where banks get involved. Nova targeted the crucial part of the process that starts at the merchant's terminal: the "acquiring" of data, including customer's name, card number, amount of purchase, etc.

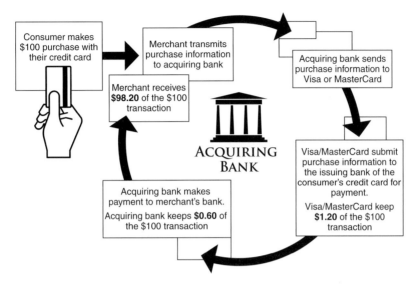

FIGURE 7.1. Credit card transaction processing showing how money flows from the consumer to the merchant. If a company like Nova does the processing, it provides the acquiring service in place of a bank, but it still uses banking relationships to manage money transfers.

In the early years of the credit card industry these data were processed by the bank that had signed up the merchant, which Figure 7.1 designates as the "acquiring bank." In the 1990s, however, this structure changed. Independent companies such as First Data Corporation and Nova Corporation increasingly supplied terminals to the merchants and provided the "acquiring service." In return they received a fee for every transaction. Banks are still involved in managing the flow of funds.

As Figure 7.1 also shows, each time a customer charges a purchase, the merchant is assessed a fee. A typical fee is 1.8% of the purchase price, but this rate can vary with the type of merchant and the kind of card used. Of course, larger merchants pay smaller fees. Part of this fee (1.2%) is passed on to the association that licenses the particular brand of card (e.g., Visa or MasterCard), and the rest (0.6%) goes to whoever processes the acquired merchant data.

In our figure the 0.6% goes to the acquiring bank. If Nova or another independent organization is doing the acquiring, it receives that 0.6%, and uses a portion of the proceeds to pay the bank that works with Nova to transfer funds. The acquirer (bank or other processor) authorizes the transaction and is responsible for any fraudulent charges.

Speed boost at the checkout counter

Nova's founding team had prepared a business plan for a new credit card processing (acquiring) company based on proprietary advances in the architecture of the processing function.

They were confident that there was room in the market for a new company that provided superior service. Nova's proposed technology would deliver two benefits. First, it would cut in half the time needed to complete a credit card transaction at the merchant checkout counter. Second, it would greatly reduce the cost of data processing.

Speed was a primary issue. As consumers we are so used to having credit card purchases approved in a matter of seconds that we forget how long it used to take. In the 1980s and 1990s it was not unusual to wait a minute or longer during peak hours for a transaction to go through, even when the store had a terminal. Sometimes the process failed, and the store clerk had to start it all over again, testing the patience of the other shoppers waiting in line at the checkout counter.

In the worst case the checkout clerk had to make a phone call to the processor's call center to complete the transaction manually. For many merchants this was as much of a disincentive to accepting credit cards as the high fees they had to pay.

Nova's proposed technology was expected not only to result in much faster processing times, thus speeding up customer checkouts, but to reduce the communications and processing costs on each transaction. The idea was that lower costs would allow Nova to service small merchants profitably.

Technology and sales risks

In analyzing this investment opportunity, we knew that we faced two major risks: the hazard of relying on a technology that had yet to be implemented, and the need to sign up new merchant customers quickly to reach profitability and keep the cash required to build the business under control. As I have stressed in earlier chapters, it is seldom as easy to successfully implement a new technology as entrepreneurs would have you believe. It is even more difficult to compete with entrenched competitors for new customers.

We ultimately concluded that the technology could be implemented within the estimated timeframe and costs. The big unknown was whether the company could acquire new customers fast enough to make a go of the business.

In November 1991 we made an initial investment in the company. We also committed to providing incremental financing over a period of 3 years to bring the company to positive operating cash flow. The team estimated that a total of $30 million would be needed until the company became profitable, an estimate that proved to be prophetic.

We committed to support the company with one major condition. In order to receive funding beyond the initial investment, Nova's management had to meet certain milestones that would demonstrate progress in growing the company. We call this arrangement an "equity line of credit." It is designed to give entrepreneurs confidence that we stand behind them as they build the business, but they must do so according to agreed-upon terms.

Nova got off to a fast start, mainly because its team was largely complete. The core technical infrastructure was put in place within 12 months, an amazingly short time.

Nova was able to build its communications network under a unique contract negotiated with LDDS, then a rapidly growing telecommunications company (which met its doom a decade later as Worldcom). Nova's architecture incorporated proprietary software residing on switches in the network backbone as well as in merchant terminals.

The approach proved to be both reliable and scalable, a big advantage for a company planning for future expansion. And reliability in this business means 100 percent uptime. No merchant can tolerate the inability to process a charge card because of service outages.

The Nova Network, as it became known, eventually grew to the point where it enabled transaction services to well over 500,000 merchants throughout the US. Because Nova's transaction and communications costs were substantially lower than the industry average, the company could make money on merchant accounts that were too small for its competitors.

Now all Nova had to do was enlist enough customers to pay its operating costs and start it on the path to profitability. Unfortunately, while the technical risks were quickly overcome, signing up individual merchants proved to be a much slower process than expected. Despite Nova's superior technology, merchants were reluctant to contract with a newcomer in the credit card industry.

Going into the second year of the venture, sales were lagging far behind forecasts, and it was evident that the company needed a new marketing strategy if it was to succeed. The company did find one which was brilliantly executed under the leadership of its CEO, Ed Grzedzinski, a co-founder of the company: acquiring merchant credit card service contracts from small regional banks and at the same time enlisting the banks as sales agents for new accounts.

This customer acquisition strategy made sense because most small banks viewed such service contracts more as an accommodation to their customers than a profitable activity. They would welcome a novel way of outsourcing what they saw as a cost center.

After canvassing the country, Nova found several small community banks willing to sell their portfolios at reasonable prices. With its first few deals the company picked up the revenues from hundreds of regional merchants. The transfer from the banks went smoothly because the company had done an excellent job of building its infrastructure.

With the success of the first acquisitions, Nova quickly switched from acquiring merchant contracts one at a time through individual sales agents to acquiring them in regional groups directly from the banks. This got the company around the reluctance of merchants to move to a new, untested provider. It also allowed Nova to win multiple merchant accounts with every successful portfolio acquisition. Merchants felt comfortable moving to a new provider because it was done with the blessing of their own bank.

In 1993 Nova amassed several thousand small merchants in this manner, thus becoming a credible player in the credit card industry. Its newfound prominence made it increasingly easy for the company to complete new acquisitions. Banks viewed a deal with Nova as an opportunity to reduce internal costs, get some cash up front, and still maintain their relationships with the merchants so they could offer them other services.

Once a portfolio acquisition was complete, Nova signed marketing contracts with the banks, effectively using them as regional sales organizations to acquire more merchant accounts. On the operations side Nova moved the data processing for the accounts from the banks' computers onto its own infrastructure, which had the benefit of providing superior service to the merchants. As the banks continued to sell the Nova service to their customers, they continued to be compensated.

Nova's strategy was producing good results, but new operational challenges confronted the company as it grew. It addressed all of them successfully:

- managing a new distribution channel: namely, the banks that dealt with and recruited local merchants;
- setting up an acquisition organization that would identify opportunities and efficiently integrate acquired portfolios into the Nova infrastructure;
- upgrading computer systems and software to service a very large number of widely dispersed merchants;

- enhancing software technology to monitor merchant charges for potential fraud, a crucial task since Nova would be responsible for all losses due to fraudulent activity. The company developed its own monitoring software, which proved so effective in spotting suspicious merchant credit card charge behavior that Nova's fraud-related losses were among the lowest in the industry.

Since the Nova management team was handling its growth very well, Warburg Pincus provided additional funds to allow the company's expansion.

Reaching the next level

By 1995 Nova had become the eleventh largest credit card processor in the US, with a sales network of 1,500 bank branches in sixteen states and a total of 77,000 merchants. Its annual revenues had reached $93 million, and the company was just beginning to be profitable.

At the same time, however, it was operating in an industry in the throes of consolidation. The top ten processors were responsible for 69 percent of the total credit card volume, leaving 400 other processors (including Nova) to fight for the rest. As the biggest processors grew in size, they were achieving economies of scale that threatened to cancel the cost advantage that Nova derived from its technology.

Nova had to find a way to rapidly increase its processing volume. It was faced with two alternatives: (1) merge with or sell itself to a consolidator; or (2) become a consolidator and enter the magic circle of the top three or four processors.

A sale or initial public offering (IPO) of the company in 1995 would have produced a reasonable return on the capital we had invested, along with good gains for the company's founders and employees based on their stock options. On the other hand, growing the company would require major investments in additional staff and technology upgrades. It would take flawless execution to pull it off, but it had the potential to produce much greater financial rewards.

After extensive discussions with the management team, we decided to go for the brass ring and continue to invest in the company's growth. As it turned out, waiting was the right decision. Within just a few months Nova was negotiating a deal with First Union National Bank (later merged into Wachovia, itself acquired by Wells Fargo in 2008) that promised to catapult it into the next league.

First Union was looking for an attractive financial option for its credit card processing business, which was about the same size as Nova's. The bank was not interested in selling its portfolio outright because it wanted to maintain contact with its client merchants, who could be customers for its other services. We offered to merge their business with Nova's and give them a large (but not controlling) equity ownership in the combined company.

It was no easy task to convince an old-line bank like First Union to trust a major piece of its business to a startup. The bank's management had concerns about providing quality service to its merchants, and it was still worried about losing some long-term relationships once it gave up control of its credit card processing. In addition, it had never dealt with venture capital or private equity firms, and it was suspicious.

Getting a commitment from the bank proved to be a slow process, but Nova had established an excellent reputation in the industry. Eventually the agreement was finalized.

We were still considering an imminent IPO. However, we decided to postpone it for at least 12 months to give us time to complete the merger and ensure the success of the consolidation. Given how fickle public markets are, this was a risky move, since the window for IPOs can close very quickly. But we felt that the combined businesses would create a great deal of increased value, making a delay worth the risk.

The two businesses merged in 1996, and the bank's credit card processing successfully moved onto Nova's technology. Virtually overnight Nova became the seventh largest processor in the US, expanding its reach into twenty-three states with a total of 4,500 bank sales branches and 83,000 merchants.

In 1996 the company completed an IPO with a listing on the New York Stock Exchange. The offering raised $58 million. Nova's revenues for the year were $520 million, with earnings before interest, taxes, depreciation, and amortization (EBITDA) of $47 million.

Nova moves up to number three

Taking on First Union's processing business was a harbinger of greater things to come, and it further increased Nova's standing in the industry. Larger banks were now interested in selling their portfolios to Nova. In 1997 the company acquired the account portfolios of several other big banks, making it the fifth largest processor in the US, with 6,000 bank branches in forty states.

Nova made more acquisitions in 1998, increasing the company's sales outlets to 8,500 bank branches in all fifty states, in addition to 1,500 sales agents who addressed the larger merchants. Revenues topped $1.1 billion with an EBITDA of $134 million.

While Nova had a well-developed process for taking on account portfolios from banks, handling the acquisition of large operating companies sometimes proved more difficult. PMT Processors, another publicly traded card processor that focused on small and medium-sized merchants, proved particularly difficult to integrate into the Nova infrastructure because of its size and poor internal systems. Nova had to invest a massive amount of effort in bringing that part of the business up to its standards.

In 1999, having been in the investment for about 8 years and achieved our objective of actively helping to build an industry leader, we decided it was time to exit. Warburg Pincus received about ten times its investment of $30 million.

It is a tribute to the quality of Nova's management team that the actual Warburg Pincus investment of $30 million exactly matched its original estimate of what it would take to make the company profitable.

The company's progress continued. In 2000, Nova's revenues reached $1.6 billion, with an EBITDA of $226 million. It had become the third largest credit card processor in the US.

U.S. Bancorp acquires Nova Corporation

By any measure Nova Corporation was a success. It had vaulted into the very top echelon of its industry, it was very profitable, and it was continuing to grow, although more slowly. It had succeeded in displacing established vendors in an existing market, overcoming barriers to entry that would normally stymie a new company.

In 2001 U.S. Bancorp expressed interest in acquiring the company. U.S. Bancorp was one of the few major banks still committed to staying in the credit card processing business, in addition to issuing the cards.

By acquiring Nova for $2.2 billion in June 2001, U.S. Bancorp created a business that processed $100 billion of card volume from 560,000 merchants throughout the United States. Ed and his team became the managers of the combined entity, a testimony to the quality of the original Nova organization.

The legacy of Nova can be seen in its successor business within U.S. Bancorp, Elavon Merchant Services. Elavon, which resulted from the merging of the Nova business with euroConex, serves as a single source for payment processing worldwide. It processes transactions from more than one million merchants around the world every day.

MAXIS: PC GAMES AS A BUSINESS

The story of Maxis shows how a unique software product can launch a successful company, and even change a whole industry. But it also highlights how important the size of such a business can be when market dynamics change. In a market that constantly demands new products of increasing sophistication, and is also going through a consolidation of sales channels, scale matters.

Background

When Maxis was founded in 1987, the world of video games was very different from what it is today. Games were mostly designed to run on game consoles, which are specialized computers like today's Nintendo Wii or Sony PlayStation. While some games also ran on personal computers, a sophisticated game that ran only on a PC was a novelty.

But that is just what the founders proposed to market. Maxis was a partnership between Jeff Braun, an entrepreneur who had just sold his startup, and Will Wright, a young developer of PC games then in his twenties. At the time Will was working on *SimCity* for PCs. *SimCity* was released for sale in 1989, and slowly attracted attention as a uniquely intellectual game.

Jeff and Will knew that they needed outside capital to build a serious business, so they started looking for investors. We would not usually consider investing in game companies, as they had been at the root of a small 1980s investment bubble that ended badly. But *SimCity* differed from the kids' action games that had defined the field to that point.

This is a highly creative game. Rather than shooting dragons, the player builds towns. When it appeared, it promised to open a totally new market for intelligent simulation games that appealed to kids and adults alike. In addition, unlike games with winners and losers, *SimCity* encouraged exploration and risk-taking, providing an enriching experience regardless of the outcome.

We decided that this would be an interesting investment. We were not alone. Maxis was receiving a lot of attention from other venture capitalists. Jeff and Will worried about losing their independence to financial investors with short-term objectives. But we convinced them that we would be good partners in building the business to the mutual advantage of all parties.

Investing in "serious fun"

The 1992 Warburg Pincus investment in Maxis was based on our belief that Will had created a totally new genre of entertainment software for the consumer market. In addition, the proliferation of less expensive PCs for the home provided the company with an expanding base of potential customers. We also thought that the company's technology could be applied to games tailored for different customer interests and ages.

Our investment allowed Maxis to add talent in key areas, the first step in building the company. Their biggest need was for designers.

Commercial game development has a lot in common with commercial animation. It calls for a studio-like strategy, with talented designers working under a creative leader in an environment that stimulates innovation. It also demands a high level of productivity. Like a book or music publisher, Maxis needed to generate a continuous flow of new products and recruiting new talent was the first priority. But we discovered that while many people loved the idea of becoming game designers, very few had the creative skills to build best-selling products. Therefore, the company licensed products from startups and made acquisitions of small companies with proven products in addition to internal development.

Furthermore, the company had to develop new domestic and international distribution channels. Its products were primarily sold in software stores such as Programs Unlimited and Electronics Boutique, which were common at the time.

Maxis achieved a good deal of success. Updated versions of *SimCity* remained best-sellers for a long time. The company also capitalized on the *Sim* franchise by releasing titles such as *SimEarth, SimLife, SimCopter, SimGolf, SimPark,* and *SimTunes*. *SimGolf* was the first Maxis sports game, later supplemented by *Tony LaRussa Baseball* and *KickOff Soccer 1997* from outside developers. Other releases included *Full Tilt 2, Pinball, Mary in Where's Morgan?,* and *SimCity 2000* for the Sony PlayStation in Europe.

Maxis completed an IPO on NASDAQ on May 25, 1995 raising $30 million.

Hit games and marketing challenges

Between 1994 and 1996 the company's annual revenues increased from $23 million to $55 million. But the game industry was changing, and the company needed experienced senior management from the entertainment industry. In 1996 we recruited Sam Pool from the Walt Disney Corporation to be CEO, with Jeff Braun moving to chairman of the board.

The management team had a worrisome combination of two strategic issues to deal with. The retail software sector was rapidly consolidating, and it was getting more expensive to create new games.

The first of these two issues had the most telling effect. Back when you could find two stores specializing in computer programs in every shopping mall, there was plenty of shelf space for games. The stores even carried niche items that did not sell many copies.

But in the mid-1990s the retail software market consolidated, and many stores closed. That meant less available shelf space. The surviving retailers required ever-increasing unit sales on every game they carried to justify keeping it in inventory.

As a result, retailers only stocked premium titles, and "hits" dominated the stores. This made it increasingly difficult for Maxis to get retail exposure for the secondary titles in its catalog. To make up for the resulting revenue shortfall the company focused on developing alternative but unproven mail order and Internet-based sales channels.

Escalating development costs for new games, the second strategic issue, resulted from the rapidly increasing processing power of PCs. It was a case of hardware driving software. Customers wanted games that made maximum use of the processing and graphic resources of their PCs, so developers had to make their programs more and more sophisticated to satisfy market demand.

Combine this trend with the need for titles that would generate large sales, and it is easy to see how product development risks had increased to an uncomfortable level. Will remained the most innovative member of the product teams, but producing a constant string of hits was increasingly difficult because of the growing effort each game required and the scarcity of extraordinary design talent.

Faced with these issues, the board was considering ways to more rapidly grow the company when it was offered the opportunity to merge with Electronic Arts. Electronic Arts was aggressively consolidating its products in both console and PC games, and Maxis was an attractive acquisition because of its reputation for innovation.

The board decided to accept the offer. In 1997 Electronic Arts acquired Maxis for stock valued at $125 million. The Warburg Pincus proceeds were approximately three times our investment of $10 million.

The Maxis legacy

This was not the end of the Maxis franchise. Its unique style survived the merger when Will Wright joined Electronic Arts. Although he had already conceived of the idea for a new line of games called *The Sims*, it would take several years of work, a lot of new technology, and a large team of multimedia developers to complete this complex product family. These resources were available at Electronic Arts; they had been out of reach for a company the size of Maxis, where such investments would have produced big operating losses.

In the *Sims* games the players create people, characters who inhabit a virtual environment also created by the players. Instead of simply building a virtual city as in *SimCity*, players build a virtual society populated by Sims. When it was eventually issued by Electronic Arts, the *Sims* series became the best-selling PC game ever.

Will was recognized as the most successful PC game designer in history.[2] So much so, in fact, that the biggest magazine devoted to PC gaming led off its review of his recently introduced game *Spore*

---

[2] D. Kusher, "Engineering Spore," *IEEE Spectrum* (September 2008), 36–40.

with the headline "The universe is your oyster in the latest entry from Will Wright."[3] Few, if any, game developers have achieved that degree of name recognition.

SYNQUEST: OVERTAKEN BY THE MARKET

While Maxis had a narrow product focus, our next case history involves a company with the much broader ambition of selling software that would drive the manufacturing supply chain of big manufacturing companies.

Entrepreneurs frequently underestimate the difficulty of selling mission-critical applications software to enterprises. The SynQuest story shows how elusive success can be for a startup, even one with good management and solid technology, when it tries to convince large enterprises to trust crucial business processes to its software.

Big ambition incurs big risks, and the story of SynQuest is a good illustration of what that can lead to. Led by an outstanding CEO, Joseph Trino, SynQuest offered excellent technology and did find many customers. Yet it failed to build a sustainable business. It could not scale up rapidly enough to achieve sustained profitability. The stars just did not align for its financial success.

Background

In the early 1990s we were looking for new areas of software with interesting investment potential. Our research convinced us that software to automate manufacturing processes was going to emerge as an important product category.

Manufacturing Resource Planning (MRP) software had blazed the trail, and by 1994 it was a $2.6 billion annual business. However, since MRP products focused on financial planning and sales forecasting, we saw an opportunity to supplement their coverage with software targeted at the actual manufacturing process.

---

[3] W. Smith, "Spore," *Maximum PC* (December 2008), 95.

There was also a technology play. MRP systems generally operated on mainframe and proprietary minicomputers. There were no software products on the market for real-time factory floor scheduling and resource planning that ran on open-architecture distributed computer systems.

From our discussions with industry analysts and potential customers, we came to believe that the market for supply chain management software would match that for MRP products. The challenge was to get into the market early enough to establish a solid beachhead, because competition was going to be fierce.

Our idea was to fund and build a software company that would offer a market-leading, comprehensive set of interrelated products. These would help manufacturers manage complex processes that were often poorly coordinated, including sales, vendor sourcing, product manufacturing, and fulfilling customer orders.

Innovative concepts, comprehensive approach

We took the first step in late 1994 when Warburg Pincus acquired FACT, a startup in Albany, NY.

FACT had developed a new class of software that used sophisticated scheduling and workflow algorithms to automate factory floor scheduling activities. It already had a number of installations in place at smaller companies, but the software needed extensive enhancements to make it suitable for large manufacturing operations.

To build a company with the scale needed to compete, in 1996 we recruited Joseph Trino, an experienced and highly respected senior executive from the manufacturing software industry. FACT was then renamed SynQuest.

Recognizing the need to enter the market with strong products as rapidly as possible, the company decided to create the planned integrated product suite through a combination of internal product development and the acquisition of externally developed software. To this end, SynQuest acquired three US-based companies and a French firm between 1995 and 1997. As a result, the company quickly expanded its market beyond the narrow specialty of factory floor manufacturing execution to position itself as a vendor of full-functioned supply chain management software.

Ultimately SynQuest introduced a software suite that synchronized production planning, scheduling, execution, and distribution across a manufacturing company. It also developed the architecture for Internet-based products, which not only delivered all the benefits of its offline product portfolio, including integrated planning and execution software and financial optimization, but also offered the wide deployment and easy access of an online solution.

Figure 7.2 illustrates the functions of the various components of SynQuest's software suite as it existed in 2000, at the time of its IPO. As you can see, the company offered a comprehensive solution to manufacturing companies. Just as important, it was flexible in operation. Users could not only plan and track each step of the manufacturing process; they could change the plan on the fly to account for unforeseen circumstances.

The software suite was designed to track an order for a manufactured product from initial sale to final delivery. Once an order was entered, the software established all the actions necessary to fulfill it, along with time frames and the interrelationships among the various functions.

Since dynamic sourcing of supplies is critical to a manufacturing operation, the software compared events in the supply chain with the original plan. Deviations from plan were used to reforecast the dates for order completion and delivery. Users could also take corrective action to get around problems, such as specifying alternative ways to get products sourced.

Obviously a supply chain management system is not software for occasional use. Once installed, it becomes a vital part of a manufacturer's operations, and users expect it to continuously evolve with changing business requirements.

A company thinking about buying such a system wants to know that the software vendor is around for the long haul. For that

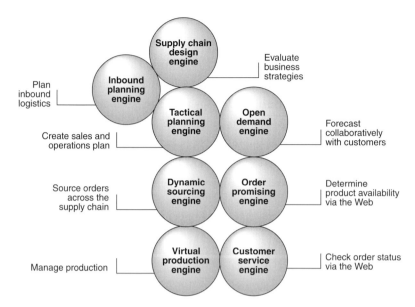

FIGURE 7.2. The supply chain management software suite marketed by SynQuest in 2000, at the time of its initial public offering.

reason a startup such as SynQuest has to work very hard to get potential customers to consider its products. That clearly makes closing orders more of a challenge, but SynQuest did have one huge positive on its side: the product suite was ahead of the competition, at least initially. Customers really had no other equivalent choice.

Even with that advantage, SynQuest needed to prove its credibility to big companies. Its marketing strategy was to go to market with distribution agreements with vendors that the potential customers already knew. J.D. Edwards and IBM certainly filled the bill, and over a period of 3 years they participated with SynQuest in selling its manufacturing suite to over fifty customers. These agreements opened doors but came at a price. The partners took their share of the revenues while SynQuest incurred the big cost of software implementation and customization to specific customer requirements.

Everyone associated with SynQuest knew how risky it was to depend on large distribution partners, but it appeared to be a necessary first step in penetrating the market. The expectation was that over time, as its products gained acceptance in the industry, SynQuest would establish enough credibility to market them directly.

SynQuest's customers, who came from a broad cross-section of manufacturing industries, were happy with the products. The software was installed at furniture companies such as Herman Miller, metal products companies like Reynolds Metals, plastic product manufacturers such as Crane Plastics, food manufacturers such as Sara Lee, and US-based automobile manufacturers like Ford Motors and Honda.

The suite's performance in customer installations was highly encouraging. For example, during 18 months of use, a division of the Herman Miller furniture company increased its on-time shipments from 87% to 99%. Its average time to complete an order dropped from 16 days to 5 days, and annual inventory turns reached fifty, a three-fold increase over the previous year.

Another example of the suite's effectiveness was the experience of the Chesapeake Display and Packaging Company, which produces displays for merchants. Chesapeake used SynQuest software to link up with its suppliers over the Internet. It reduced manufacturing cycle times by 50%.

While SynQuest was making progress, however, the competition was becoming increasingly intense. Large companies like SAP were entering all sectors of the supply chain management market, increasing the effort SynQuest needed to acquire new customers.

Then SynQuest's primary distribution partner, J.D. Edwards, decided to expand beyond its MRP roots and enter the supply chain management field with products of its own. Having lost a vital ally, SynQuest began a search for other sales channels, and expanded its own sales and marketing efforts to fill the gap.

A public offering but no profits
Building the company had consumed a great deal of capital. To raise more, SynQuest launched an IPO in August of 2000. At that point the company had annual revenues of about $25 million. Its IPO on

NASDAQ raised $32.6 million on the strength of its revenue growth and market prospects.

Initially the company's stock was well received. It had been issued at $5 a share and reached $18 just 2 months later, in October. That gave the company a market value of about $400 million.

But while the company was making progress in its stock price and its product deployments, it was posting continuing operating losses due to the rising cost of sales, customer support, and product development. Sales costs were especially onerous. Because SynQuest's complex products were being sold into an increasingly competitive market, it took more and more effort (and money) to acquire each new customer.

Overall, the revenues of the company grew from $1.6 million in 1995 to $32 million in 2001. But its operating losses grew too, reaching $13 million in 2001.

As the board analyzed ways to reach profitability, it became clear that two strategic challenges had to be assessed in deciding whether to invest more capital. First, the company would continue to incur large product development costs because the market required rapid product evolutions. Second, as manufacturing companies left the US and the market for manufacturing software went global, SynQuest would have to initiate an international sales effort. The expenses involved in sales and product implementation outside the US was going to cause costs to escalate still higher.

To make the situation even more difficult, potential customers understandably placed considerable emphasis on the financial viability of their vendors of mission-critical software. As a public company, SynQuest's thin financial resources were public knowledge. Industry analysts did not fail to point this out in their product advisory reports – "great products but weak finances." This hurt sales growth, particularly as competitive products emerged from big, financially solid competitors.

A few years earlier it might have been feasible to solve the problem by raising more capital, but after the 2000–2001 stock market crash that road was closed. The board decided that a sale of the company was the best way of preserving value.

At the end of 2002, SynQuest merged with Viewlocity, a company with complementary products. Unfortunately, as the supply chain software industry consolidated into a small number of large vendors, Viewlocity became a marginal player. Warburg Pincus had no return from this investment.

LESSONS LEARNED

All three of these case histories highlight the importance of successful marketing strategies and company scale to complement technological excellence. They also demonstrate the business challenges faced by management teams as they try to cope with the incredible pace of change in the software market.

It is expensive to sell enterprise-level software
to big corporations

The story of SynQuest illustrates the challenges of building a company to sell mission-critical software to big enterprises. Customer willingness to buy hinges not only on superior technology, but also on the perceived financial stability and established reputation of the vendor.

Cracking that barrier requires very unusual circumstances: a combination of a big lead on big competitors *and* a value proposition that is unbeatable in terms of rapid return on investment for customers. It also requires enough capital to patiently build a business of sufficient size to be profitable. And dependable marketing partners are a must.

Customers have good reasons for their caution. Mission-critical software, such as SynQuest's supply chain management system, is totally different from the "shrink-wrapped" programs used in many consumer applications.

In addition, no two manufacturing companies are alike. For customers to derive full benefit from SynQuest's software, it had to address their different and highly specific manufacturing issues. SynQuest accommodated the needs of its customers, but it could not do so profitably at its size. It had to grow much more rapidly to reach the right enterprise scale for profitable operations.

SynQuest identified and addressed an emerging market early on, before entrenched competitors with far greater financial resources gained control. Unfortunately, it did not have enough of a head start to build its business before the big players arrived and the center of gravity of many manufacturing companies shifted off-shore.

As a result, the size of investment needed for it to sustain a leading position skyrocketed, and SynQuest simply did not have the financial resources to match what other, larger, more established companies brought to the market. Had the public markets continued to be favorable for raising more capital, the story might have ended happily.

Managing acquisitions is a crucial task
Very few companies in fast-moving technology markets can afford to rely exclusively on internal product development for their growth. All three companies discussed in this chapter made acquisitions that were essential for their survival and development. Nova's acquisition strategy, in particular, was central to its success.

First, Nova's management was very good at identifying candidates for acquisition and negotiating the price. Nova also had a dedicated integration team that quickly consolidated operations onto its platform. This allowed the company to grow the business without the problems and costs of running dual systems for an indefinite period of time. In fact, the company integrated more than sixty acquisitions within a few years.

Maxis also acquired a number of products and companies, but none of the game developers in those companies produced the rare "hits" that the market demanded.

SynQuest's highly experienced management developed an acquisition strategy that succeeded in complementing the company's internally developed products with externally produced software, enabling the company to offer a comprehensive product suite in a relatively short timeframe.

A winning marketing strategy is just as important as good technology

A common misconception of entrepreneurs is that the best technology wins in the market. This may be true eventually, but getting the attention of willing customers comes first. Developing a winning marketing strategy requires just as much creativity as product development.

Nova would not have survived had it stuck to its original idea of selling its service directly to merchants. While the company had superior technology from the start, it was the development of new distribution channels through regional banks that made it successful.

SynQuest depended on big corporate partners for its distribution partners. Such relationships are fickle and the company was not able to find a low-cost marketing strategy when its most important partner decided to sell its own competing products.

Going public has its pitfalls

All three of the companies discussed here had public offerings and raised capital under highly favorable conditions. Entrepreneurs have always considered an IPO as a badge of success. It usually is, but there are potential problems when business conditions become difficult and the company's share price drops. A low or declining stock price sends a negative image to customers and makes employees unhappy.

SynQuest was not profitable, and attracting capital after the stock market bust of 2000–2001 proved impossible. Furthermore, the company's finances were public. With operating losses in plain view, customers and competitors found it easy to question its viability as a vendor. All these grief factors combined to make a recovery much more difficult than it would have been for a private company.

In the case of Nova, the company was already profitable, so the capital was not used to fund losses. That helped public perception. The same was true of Maxis.

Scale matters

In Chapter 3 I highlighted the consolidation process that occurs in practically all industries, but particularly in software. The three companies we have discussed here are very good examples of how critical scale is in building a sustainable business in a highly competitive market. You need to get big enough to stay in the race, and it has to happen fast. The problem is that building scale in a startup is especially costly because investments are made ahead of revenues.

In the case of Nova, the challenge was to create a revenue base large enough to support a costly infrastructure. The success of the company was ultimately the result of a creative customer acquisition strategy supported by outstanding management and technology. Revenue growth was fast enough that the amount of invested capital stayed modest.

For Maxis, scale meant a larger pool of game developers and a revenue base that would enable more efficient distribution strategies. Generating "hit" games is, to a large extent, a statistical process that requires a broadly diversified talent pool and an extensive product line. As Maxis grew we realized that the company either had to acquire the additional talent and products or be acquired itself by an industry consolidator. This was a major factor in the decision to merge the company with Electronic Arts.

Finally, the situation of SynQuest shows what happens when a company cannot reach critical mass because of funding limitations. While the company won a significant number of happy customers, its revenues were simply too small for it to be profitable in the face of mounting competition while also continuing to expand its products and customer base.

CONCLUDING COMMENTS

As long as there are computers and people to use them, new software companies will spring up to create programs for business and consumers. Since software is almost pure intellectual property – computer code written by experts to perform a specific task – its forms and uses are limited only by the human imagination.

If you require proof of the infinite potential of software, just look at how many new possibilities for software applications and how many business models for delivering software functionality have been opened up by the Internet.

One of the new distribution models enabled by the Internet is "software as a service" (SaaS). In this model, the software is hosted and maintained by its vendor. Customers access and use the applications through the Internet. They pay for the software on a monthly schedule, based on usage, rather than up front as with a traditional license purchase.

Another advantage of SaaS is that customers are freed from the need to manage the software on their servers, with all the support costs that this entails. Users automatically have the latest version whenever they log into the application. Salesforce.com is the most visible example of this model, and its success has prompted even some large, successful vendors of licensed software to follow its lead.

Software will continue to generate business opportunities for the foreseeable future, as clever people constantly build innovative applications to serve newly discovered market niches. It also helps that the perceived investment cost is deceptively low.

What does this mean for the venture investor? First, there will be lots of chances to make a bet on a promising company. Second, the investor will have a wide choice of markets.

The three startups we have analyzed in this chapter underscore this second point. They addressed completely different markets and developed highly dissimilar kinds of software. One was a developer of computer games for consumers. Another was a financial services company that built its business around proprietary software. The third was a software products company marketing to large enterprises.

All three companies had superior products. But as we have seen, product quality is only one of the requirements for business success.

Investors in a new software company, or a company basing its business on proprietary software, need to look closely at its market and the scale needed for profitability. Most important, they must gauge the time needed for competitors to attack the company's lead in the market.

What must we do next to stay ahead? Few entrepreneurs have a good answer to this question, but the ones who do deserve encouragement.

8 Venture capital: Past and future

> The advancement of the arts, from year to year, taxes our credibility and seems to presage the arrival of that period when human improvement must end.[1]

Henry L. Ellsworth, Commissioner of the Patent Office, 1843

No, we have not seen the end of industrial innovations. In fact, the pace of technology development continues to accelerate, and the world depends on innovations to drive economic growth. New technologies continue to appear, promising to improve human health, accelerate business processes, open new avenues for renewable energy generation, or accomplish any number of other worthwhile missions.

Without risk capital many of the innovations of the past 40 years would have had a negligible impact on the economy. Throughout this book we have seen the crucial role that venture capital has played, particularly in the US, in funding innovative young companies that create new markets and new industries with their technology.

But the story is not wholly positive. For two decades, between 1980 and 2000, venture capital availability increased at a remarkable rate, giving venture capital firms (VCs) the means to finance all those startups. Then, with the bursting of the Internet and telecommunications investment bubbles, the environment changed. The capital available for investment by VCs declined from its peak of $104 billion in that year, as shown in Figure 2.1 (p. 44), to only $28 billion in 2008.

To make matters worse, during the same period there has been a marked reduction in the average financial returns of venture capital funds. This is a central concern of this chapter.

---

[1] Quoted in S. Sass, "A patently false patent myth," *The Skeptical Inquirer*, 13 (Spring, 1989), 310–313.

Such a downturn naturally leads to questions about the future of the venture capital investment model.[2] Is it still competitive in a changing global economic landscape? Are we at the end of the era when new companies can jump-start a whole new sector of the economy? Will the lower profitability of venture capital investing continue to attract new funds?

These are difficult questions to answer, especially given that the venture capital industry is so young. It is hard to form any solid projections for the future based on so short a history.

Personally, I believe that the venture capital model is here to stay, and that it will continue to fill an important role in funding innovative companies while generating attractive financial returns for investors. However, venture capitalists will have to become more selective in their investments, if only because there will be less capital to be deployed than there was in the peak years. VCs will also reduce the number of investments they make in a given year, because not only will there be less money to invest but it will be more expensive to build a viable business. This is the inevitable outcome of the challenges posed to young companies by the realities of a global economy and industry consolidation.

But prophecy is a perilous undertaking, especially in a world where yesterday's eternal economic verities have become today's exploded myths. So this chapter is not an exercise in crystal ball gazing, but a review of the forces that have influenced the success of venture capital investing, and of the changes that have occurred since 2000. Hopefully such insights can guide our thinking about its future.

So far we have concentrated mostly on the operational and market factors that influence the success or failure of individual investments. These included quality of management, competitive offerings, size of market, and degree of market consolidation, to

---

[2] R. Waters, "Venture capitalists face broken system," *Financial Times* (April 30, 2009), 15.

name a few. Now we will look more closely at how external factors impact the profitability of investments, with particular attention to the role that public markets play.

1980–2000 AND BEYOND

As we ponder the future, we must keep reminding ourselves that the success of venture investing is not determined by the availability of great new technologies alone. It is strongly influenced by business cycles, global trade, public capital markets, and government initiatives.

The years since 1980 provide a striking background against which to consider the challenges facing venture capital investing in changing economic environments. Looking back, investors were spoiled by the extraordinary 1980–2000 period. It was an era that offered an intoxicating combination of technological breakthroughs, a favorable economic climate, and receptive financial markets. It encompassed the greatest bull market in the history of capitalism.

The Warburg Pincus-funded companies we have tracked from initial investment to divestiture represent a cross-section of this era. They operated in five major industrial sectors: communications, semiconductors, the Internet, software, and financial services. All but one of them started as fledgling businesses. Nine became publicly traded. Not all were successes, but the blended internal rate of return to our funds for the twelve companies that we did exit was 139 percent.

That figure represents an outstanding profit profile, which is the reason VCs assume the risks they do. We will discuss the profitability of venture capital funds in some detail later in this chapter. But first we need to lay the groundwork by summarizing the major factors that created that period's extraordinarily promising environment for technology company growth. Some of these conditions have been mentioned in earlier chapters, but it is worth another look to see how they impact venture capital investing and its profitability.

What made the 1980–2000 period such a favorable one? These are the most important elements.

- *Healthy gross domestic product (GDP) growth over extended periods of time.* Rapid industrial development in the US and around the globe created economic and market conditions that encouraged the growth of new ventures.
- *Deregulation of telecommunications.* Starting with the breakup of the AT&T monopoly in 1984, new business opportunities opened up in the US telecommunications market for startups to provide equipment, software, and services to consumers and industry.
- *Growth of global markets.* Massive industrial development around the world opened new markets for sophisticated industrial and consumer products.
- *Major technological innovations reaching maturity.* Many new synergistic innovations enabled new markets.
- *Favorable financial markets.* Periodically, vibrant public financial markets placed high values on technology businesses. They became receptive to stock offerings from these new, high-growth companies, even ones that were not profitable.
- *New leaders in consolidating markets.* As these new markets matured, successful startups vied for dominance against established companies.

Let's consider each of these factors in turn.

The economic environment

Figure 8.1 shows annual GDP growth in the US from 1980 through the third quarter of 2007.[3] Although this long period was punctuated by recessions, marked by declining GDP growth, there were extended periods of strong gains.

As the chart makes clear, from 1993 to 2000 real US GDP (adjusted for inflation) grew at a rate of 4% annually. This is 46% faster than the average of the previous 20 years. That is significant for the success of venture capital investing. Periods of good GDP

---

[3] "Economic report of the President," *The annual report of the Council of Economic Advisors* (Washington, DC: United States Government Printing Office, January 2001), p. 19.

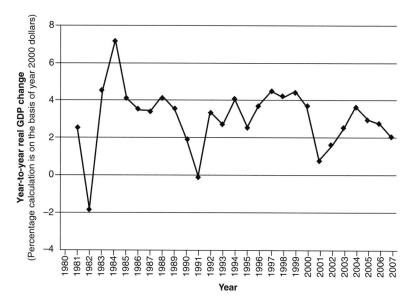

FIGURE 8.1. Year-to-year GDP change in the US; values adjusted to year 2000 dollars. *Source:* US Department of Commerce, Bureau of Economic Analysis (ref. 3).

growth provide a favorable environment for the creation and development of new companies, so it is no surprise that venture activity accelerated during this period. Between 2001 and 2007, by contrast, GDP growth slowed to an average rate of only 2.3%, and venture capital investments dropped.

Deregulation of telecommunications

In Chapter 4 we saw how the deregulation of the telecommunications industry, starting with the breakup of the AT&T monopoly in 1984 and concluding with the Telecommunications Act of 1996, brought hundreds of new companies into the US market. In fact, deregulation of telecommunications became a worldwide phenomenon, creating opportunities for startups in Europe and Asia as well.

These new companies created hundreds of billions of dollars of value as they went public or were acquired. Unfortunately, after the stock market corrections of 2000–2001, many of the startups and companies that had done initial public offerings (IPOs) disappointed their investors.

Of course, some venture capital funds benefited from their ability to exit their investments at high valuations, and these funds reported high profits. This was the case with Covad Communications. There were also other infrastructure services startups such as NeuStar that emerged as successful companies, thanks to business models with strong and enduring competitive advantages. Both of these examples were discussed in Chapter 4.

But too many other startups of that era could not survive in the face of intense competition and lack of adequate funding. For example, few startups offering local telephony services (CLECs) survived after 2001. Their high failure rate is not surprising. Telecommunications is a very capital-intensive industry, and the telephone giants could continue pouring money into building a modern infrastructure, whereas startups were starved for capital.

We see the results today. In 2009 three former regional Bell operating companies account for practically all of the wireline service in the US. All three are big and well capitalized. The competition didn't stop there. Cable TV operators leveraged their extensive infrastructure and massive cash flow to enter the telephony and broadband service markets, too. These giants relegated new entrants looking for a solid foothold to niche markets.

A similar pattern prevailed in the Internet market, the twin of the telecommunications arena. Very few of the thousands of Internet startups of that era prospered by developing profitable business models, as discussed in Chapter 6. Most fell victim to excessive competition and impractical business models.

When the telecommunications and dot-com bubbles collapsed after 2000, the failures of so many 1990s startups badly impacted the profitability of venture capital funds that had invested heavily in such companies.

Growth across borders

Rapidly increasing global economic development opened new world markets for technology companies. This represented both an opportunity and a challenge to venture capital-backed startups.

The opportunity lay in the sale of high-technology products into emerging international markets. The challenge was to attract enough investment capital to equip companies to operate on a global level.

That is not an easy problem to solve. International operations cost a lot more to run than regional or national ones. Just as companies in the telecommunications field need lots of capital to compete, so global enterprises must have sufficient scale to build sustainable businesses.

When a global strategy is successful, the rewards can be substantial. We saw in Chapter 5 how Zilog built its business model around sales to fast-growing consumer electronics companies in Asia.

But after 2000 this changed, as competition intensified and the cost of doing international business grew. Asian companies emerged to fill local requirements, and soon began competing with established providers, not just in their home countries, but worldwide.

Major technological innovations create new markets

In 1980, when economic conditions were favorable to new business formation, a critical mass of new technologies stood ready for commercialization. The innovations that enabled so many startups were the outcome of years of research and development (R&D) by public and private organizations. The numbers tell the story.

As Figure 8.2 shows, annual spending on R&D in the US increased from $25 billion in 1954 to $240 billion in 1999 (in constant 1996 dollars). Note in particular the sharp increase in total spending starting in 1976.[4]

---

[4] *Ibid.*, p. 111.

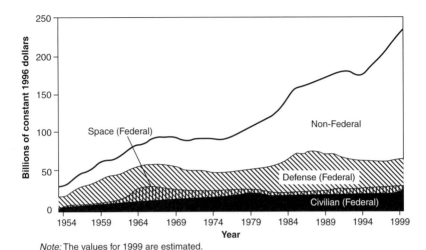

FIGURE 8.2. Federal and non-Federal R&D spending, 1954–1999.
Source: National Science Foundation (ref. 4).

This investment in R&D produced enormous advances in such fields as integrated circuits and microprocessors, fiber optic communications, wireless communications, computers, software, solid state imagers, and the Internet. These technologies formed the base from which entrepreneurs were able to launch businesses in the 1980s and 1990s.

There was plenty of talent available to take products to market. Big companies with large internal R&D budgets were often slow to harness the deep technical skills of researchers who worked in these new technologies. Startups were able to lure frustrated researchers away from corporations by offering them the chance to build businesses around technologies they had developed.[5] We saw in Chapter 4 how Epitaxx, which used technology originally developed at RCA Laboratories, was launched by scientists who had participated in its development.

Information technology (IT) is an especially striking example of how entrepreneurs and technologists drew on developing

[5] For a more detailed discussion see H. Kressel with T. V. Lento, *Competing for the future: How digital innovations are changing the world* (Cambridge: Cambridge University Press, 2007), pp. 110–120.

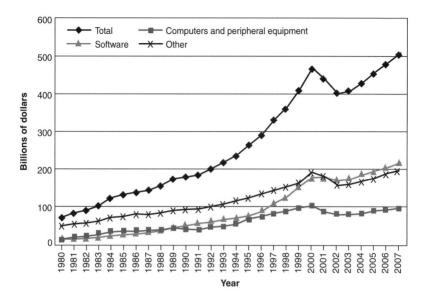

FIGURE 8.3. Information technology products revenues (industrial applications) in the US, 1980–2007. *Source:* US Department of Commerce, Bureau of Economic Analysis (ref. 6).

technology to create burgeoning new markets. With computerization driving technical and productivity gains across most industries, sales of IT hardware and software displayed remarkable growth for many years.

Between 1980 and 2000, when the amount of venture capital was growing, the total US market for computer hardware and software for industrial applications grew from $68 billion to $468 billion (Figure 8.3).[6] That amounts to an average annual growth rate of about 10 percent.

What began as computers and software soon spun out new market segments. Companies like Level One Communications, discussed in Chapter 5, took advantage of the proliferation of personal computers to launch chips aimed at the need for networking. When

[6] "Economic report of the President," *The annual report of the Council of Economic Advisors* (Washington, DC: United States Government Printing Office, February 2008), p. 248.

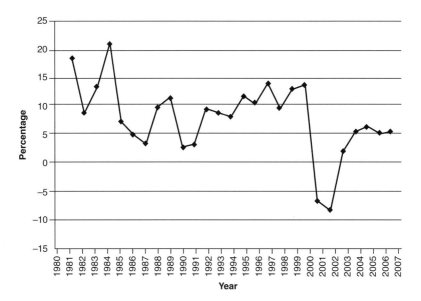

FIGURE 8.4. US revenue growth rate for information technology products for industrial applications, 1981–2007. *Source:* US Department of Commerce, Bureau of Economic Analysis (ref. 7).

its products made it possible to easily deploy high-speed data networks over existing copper phone lines, success was assured.

After 2000, as Figure 8.4 shows,[7] life became more difficult for young IT companies. They had to compete in a market that was growing only half as fast.

The picture doesn't change very much if you exclude the hardware side of information technology. After 2000 the annual growth rate of the software industry also declined, from 16% to about 5% (Figure 8.5).[8]

New industry leaders emerge as markets consolidate
Another distinguishing characteristic of the pre-2000 growth in technology markets was the number of leading companies that emerged. The increased availability of venture financing, coupled

[7] *Ibid.,* p. 248.
[8] *Ibid.,* p. 248.

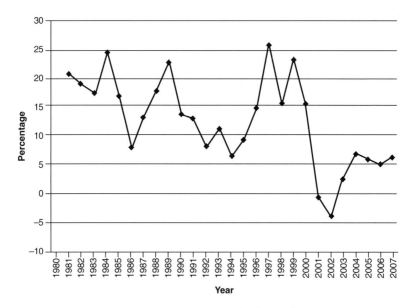

FIGURE 8.5. US revenue growth rate of software for industrial applications, 1981–2007. *Source:* US Department of Commerce, Bureau of Economic Analysis (ref. 8).

with the advent of transformational technology, opened the door for many venture capital-backed companies. A few of them became industry leaders, producing huge profits for their investors.

These startups were well positioned to take the lead in markets that had not yet consolidated – the same strategy used by Level One Communications in the networking chip arena. Several of the more aggressive upstarts grew to become large public companies and a few, such as Cisco Systems, Oracle, and Microsoft became the dominant players in huge new markets.

Executives of these new firms recognized that technology was moving too fast for them to depend only on internal resources for new product development. They were very focused on maintaining technological leadership through savvy acquisition strategies. Industry consolidation was under way in earnest.

This was a source of grief for many other startups. As well-managed companies achieved industry dominance in product areas popular with venture capital investors, they ratcheted up the competitive pressure, reducing the profitability of newer companies attracted to the industry by its high growth rate.

When investors in such companies evaluated their position, they were led to the conclusion that selling the companies was a wiser exit option than trying to pursue an IPO. As a result, many companies with leading-edge technologies were sold to the industry consolidators at attractive valuations, providing their venture capital investors with profitable exits. Laggards found themselves in a precarious position.

Big, successful new companies were not the only ones helping the process of consolidation. Even old-line companies like Lucent, Nortel, and IBM eventually became aggressive acquirers of young companies with advanced technology products. IBM alone acquired forty-six software companies between 2003 and 2008 for a total value of $14 billion.

Public market valuations and initial public offerings

Of the five major influences on venture capital profitability under discussion – GDP growth, industry deregulation, globalization, technology advances, and favorable financial markets – it is the financial markets that have the most immediate effect, and are the least predictable.

Public markets influence the valuations for the companies that VCs fund and later divest. They also determine whether the companies can access capital through an IPO. Because selling stock to the public is an important investor exit path, the state of the public markets is a key determinant of the profitability of venture capital funds.

Young companies get access to public markets only periodically. The ups and downs of general business conditions are partly to blame, but another reason is that investors in public company stocks are prone to cycles of love and hate for certain industry sectors.

When technology businesses grew rapidly during the 1980s and 1990s, followers of the public markets took notice. Stockbrokers, investment bankers, and investors alike came to love these young entrepreneurial firms. As we will see later, their enthusiasm was reflected in valuations for technology stocks in the late 1990s that were unusually high relative to historical ones.

During such "hot" market periods public interest in new technology businesses can be so high that young companies with a record of fast growth are able to launch successful public offerings even if they have yet to make a profit. Investors value such companies by looking at their growth rate and projecting future growth, leaving current profits out of the equation. But periods where this can happen tend to be rare. Normally, both growth rate and profitability are important factors in valuation.

One common valuation metric, usually applied to profitable companies, is a multiple of net income (profit), or earnings before interest, taxes, depreciation, and amortization (EBITDA). EBITDA is a measure of the cash generated from a company's operations before any reinvestments in capital equipment or facilities.

Another approach is to value a company on a multiple of its annual revenues. This metric is typically used for companies that are growing very fast but have only modest profits.

Public market valuations are not a science. Professional investors would love to discover the secret formulas that drive valuations, but unfortunately they do not exist. Valuations are highly subjective, with investors trying to anticipate both industry and company growth and their effect on the company's future profitability. One of the considerations is the perceived ability of the company to sustain its competitive advantage. A unique competitive edge can give a company "scarcity value" among investors, resulting in valuations much higher than other metrics would indicate.

If the market is receptive and its valuation of the company is positive, going public opens a path for investors to exit their investment over time. It is a powerful incentive for entrepreneurs to eventually cash in on their options. Public capital also gives a company access to resources far in excess of those offered by venture capital investors for continuing to build the business. And a publicly traded stock can be an attractive currency for effecting acquisitions.

Raising capital through an IPO can move a company into the big leagues. Covad Communications benefited from this strategy to finance its infrastructure prior to 2001. NeuStar and Nova Corporation similarly benefited from their access to public market capital.

But as we have seen, going public has its shortfalls. Publicly traded companies have to bear increased costs for financial reporting. The public scrutiny can be quite uncomfortable, and can harm the company's chances of success, as it did with SynQuest (Chapter 7). Nevertheless, most companies that have the opportunity to launch an IPO do so.

Once a company decides to pursue an IPO, it has to take into account the fact that its valuation will be very much influenced by the valuation of comparable public companies.

The vagaries of technology company valuations are evident from historical data. We will use a *multiple of revenues* of companies listed on the US public markets as our valuation measurement.

This is a reasonable metric for our purpose, because it matches the typical pattern of many of the young venture capital-backed companies. They often grow very rapidly, but without generating significant profits. In a private sale or IPO scenario, therefore, they are mainly valued based on their revenues and their growth rate, the uniqueness of their technology or products, and their prospects for future profitability.

Figure 8.6 shows the *average* ratio of market value to company revenues for a cross-section of publicly traded electronics companies from 1994 to 2008. These are large enterprises, all listed in the S&P 500.

Some interesting facts about the favorite industries of venture capital investors can be drawn from this chart.

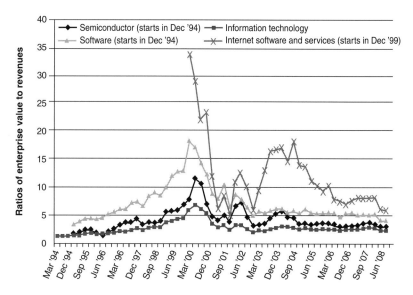

FIGURE 8.6. Ratio of enterprise values to sales for four market sectors. These data are for companies in the S&P 500 group of publicly traded companies, 1994–2008. *Source:* Based on data from Standard & Poor's.

- Semiconductor company valuations increased from 2X to 12X revenues between December 1994 and March 2000, then dropped back to between 3X and 5X.
- Software company valuations increased from 3X to 13X and then dropped back to about 5X in the same two time periods.
- Internet company valuations peaked at 34X before 2000, but have swung wildly between 5X and about 17X since then.
- Valuations for IT companies (a broad grouping of companies in the electronic equipment, software, and services sectors) went from 1X to a peak of 7X before dropping back to the 2X–3X range after 2000.

What does this mean in terms of the exit opportunities for venture capital-funded companies?

Figure 8.7 shows two curves. One tracks IPOs for venture capital-backed companies, while the second tracks the acquisitions of such companies since 1980. Note the continued high number of acquisitions versus the steep drop in IPOs after 2001. It is reasonable to assume that the two curves are related: companies that could not go public were enticed by opportunities to be acquired instead.

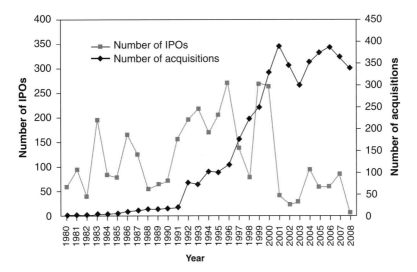

FIGURE 8.7. Initial public offerings and acquisitions of venture capital-backed companies, 1980–2008. *Source:* Based on data from Thomson Reuters.

A comparison of the data in Figure 8.6 and Figure 8.7 suggests that the active period for IPOs in the 1990s largely coincided with a period of high public market valuations. We are now ready to look at the huge impact that these public market valuations have on the profitability of venture capital funds.

VENTURE CAPITAL FUND PROFITABILITY

As we have noted several times before, venture capital investing carries a high level of risk. The statistics are revealing. Only 29 percent of the 24,635 companies funded between 1980 and 2007 returned any capital to their investors. Of these, 3,369 had public offerings and 3,857 were reported sold. The fate of the others is unknown – they most probably went out of business.[9] Nevertheless, the profits made on the winners more than made up for the losing investments. They made many venture capital funds very profitable for a long time.

[9] Data collected from "NVCA 2008 yearbook," prepared by Thomson Financial, and "PricewaterhouseCoopers/National Venture Capital Association report," from data provided by Thomson Reuters.

So the risk of investment was justified, at least on average. The question is, to what extent will this continue to be true? As far as the future of venture capital investing goes, this is the crux of the matter: just how profitable will it be? Venture capital investing has to outperform other investment vehicles in spite of economic conditions, market transformations, and financial cycles if it is to justify its high risk and attract fresh capital.

The venture model certainly seemed to meet that goal in its early days. In the case of Digital Equipment Company, for example, a total of $70,000 invested in 1957 by American Research and Development Corporation returned $355 million to the investors in 1971.

This and a few other spectacular successes made institutions with cash to invest eager to commit capital to the brand-new electronics market through professional investment firms. Attracted by the prospect of big returns, they drove the amount of venture capital available for investment from a trickle of $600 million in 1979 to a flood of $104.3 billion in 2000 (see Figure 2.1, p. 44).

Compared to their counterparts in the pubic markets, investors in venture capital funds did very well prior to 2000. Figure 8.8 tracks returns from all venture capital funds, the S&P 500, and the NASDAQ market using a rolling average of 5-year increments for the calculation. From 1990 through 2003 venture capital investments mostly outperformed the two public vehicles, often by a wide margin.

However, during the 2 years ending in 2004 and 2005, on the basis of this metric the US venture capital industry lost money *as a whole*. Not only that, it did not distinguish itself compared to the hard-hit S&P 500 and NASDAQ aggregates from 2004 right through 2007.[10]

---

[10] The point of this comparison is to focus on a trend, but it may be somewhat misleading, since the figures it uses for private equity are internal rates of return (IRRs), which are money-weighted calculations, while the public market index figures are geometric mean or time-weighted returns. Nevertheless, it provides a generalized view of the results.

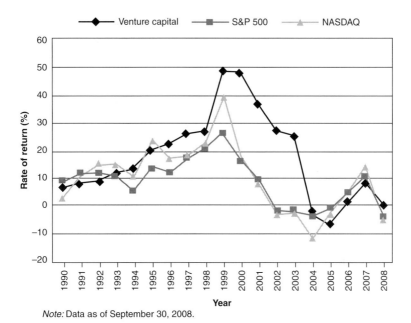

Note: Data as of September 30, 2008.

FIGURE 8.8. Five-year rolling averages rate of return for all venture capital funds and public market stocks as measured in the year indicated. *Source:* Based on data from Thomson Reuters.

During this period the public market for IPOs for unprofitable companies declined, as did overall valuations of technology companies. This subject has been extensively studied by Kaplan and Schoer[11] and McKenzie and Janeway.[12] McKenzie and Janeway conclude from an extensive historical analysis of a selected group of venture capital funds that funds that exit during the most favorable public markets have median internal rates of return of 76%, whereas those that exit in the least favorable markets have returns of only 9%. By "favorable public market" we mean a market where unprofitable companies can be taken public at high valuations, as happened in the late 1990s.

[11] S. N. Kaplan and A. Schoer, "Private equity performance: Return persistence and cash flows," *Journal of Finance*, 60 (4) (2007), 1791–1823.
[12] M. D. McKenzie and W. H. Janeway, "Venture capital fund performance and the IPO market," unpublished Working Paper 30 (2008), Center for Financial Analysis and Policy, University of Cambridge.

A matter of timing

Earlier we reviewed the developments that impacted venture capital investments before and after 2000. VCs that can exit their holdings through public offerings or sales in periods of high valuations can obviously generate high rates of return for their funds.

On the other hand, when IPOs are not possible, valuations are depressed, and it is difficult to sell companies, VCs need to hold their investments in companies longer and even continue to provide capital to finance losses. Tying up capital in the same companies can greatly reduce the internal rate of return of the funds that own them. That was the case after 2000–2001, when the decline in Internet and telecommunications stock valuations made IPOs difficult, if not impossible, for companies in the affected sectors.

Not all venture capital investors are equally affected by economic downturns. While the profitability of all reporting venture capital funds showed a marked deterioration after 2000, the top-tier VCs did considerably better than the average. This is seen in Table 8.1 which shows the internal rate of return of funds started in various years beginning in 1969.

Two important factors impact venture capital profitability. First is portfolio timing. Since a fund is usually fully invested by year six or seven, and mostly liquidated by its tenth year, its profitability is closely tied to when it was started and when it was liquidated.

It is easy to understand the importance of timing. For example, imagine an investment made in an Internet or semiconductor company in 1999, when valuations of technology firms were at their peak. Then think what the financial outcome of this investment would be if the company had to be sold in 2005, after the sharp overall drop in market valuations. You would expect the returns to be lower than those for a comparable company sold in, say, 2000, and you'd be right.

The second factor has to do with the ability of the firm's partners to attract high quality entrepreneurs and contribute to the successful outcomes of their investments. The two papers referred to

Table 8.1: *The internal rate of return (IRR) for venture capital funds started in the years shown, calculated for several different metrics.*

Vintage year	Number	Average	Capital weighted average	Pooled average	Maximum	Upper 25%	Medium	Lower 25%	Minimum
1969–75	13	19.30	18.80	19.80	36.20	24.50	19.90	12.50	7.80
1976–79	17	31.40	30.40	29.30	74.10	45.70	34.30	14.30	3.20
1980	18	13.50	22.30	18.70	31.80	18.20	13.40	8.80	-1.90
1981	22	8.00	10.40	10.90	25.40	13.60	9.90	0.10	-3.30
1982	28	2.70	4.50	4.80	13.50	8.80	3.90	-1.30	-21.40
1983	59	5.20	6.60	8.00	41.50	9.30	5.00	0.70	-57.50
1984	66	4.90	6.00	6.00	25.50	11.30	3.50	0.90	-18.40
1985	46	8.20	9.20	10.00	28.30	15.40	8.80	2.40	-41.50
1986	43	7.00	10.10	12.50	25.60	11.10	6.20	2.20	-11.20
1987	63	7.50	13.50	14.70	31.50	17.60	7.90	-0.30	-37.80
1988	44	12.30	19.80	20.40	42.70	21.00	9.50	1.60	-10.20
1989	53	12.10	16.30	17.80	55.90	17.10	10.90	0.80	-35.90
1990	22	17.50	24.40	28.20	74.90	29.30	14.00	-0.10	-8.80
1991	18	18.70	28.90	28.40	61.30	25.60	17.80	4.40	-0.70
1992	27	26.10	29.40	35.50	116.30	38.90	13.90	11.20	-47.20
1993	41	21.60	28.60	37.40	98.60	39.10	12.60	0.70	-25.00
1994	39	25.40	33.10	39.30	112.90	41.00	15.60	3.40	-47.90

Table 8.1: (cont.)

Vintage year	Number	Average	Capital weighted average	Pooled average	Maximum	Upper 25%	Medium	Lower 25%	Minimum
1995	49	46.10	57.30	59.80	247.80	64.90	21.20	3.30	-28.90
1996	35	74.30	59.20	83.30	454.90	113.90	29.70	1.40	-16.60
1997	62	49.30	46.10	49.50	296.00	59.70	20.10	-0.70	-28.70
1998	77	26.00	24.00	18.40	721.00	11.00	1.80	-3.90	-44.80
1999	110	-4.90	-6.70	-5.50	140.00	0.80	-6.20	-14.60	-100.00
2000	124	-1.70	0.40	1.40	28.40	2.70	-2.30	-7.10	-24.80
2001	57	3.20	3.20	4.30	28.00	9.60	1.90	-3.90	-24.40
2002	20	1.60	1.80	3.80	26.20	3.80	-1.10	-2.30	-10.70
2003	17	6.10	6.90	6.90	25.10	7.30	4.80	2.10	-7.50
2004	23	2.80	4.50	4.30	24.00	12.10	0.80	-5.80	-11.00
2005	17	3.90	3.70	7.30	24.10	7.30	5.50	-1.70	-20.20

Notes: Report date as of September, 30 2008 – venture capital funds only

Calculation type: IRR. Primary market: US

Source: Thomson Reuters

above (references 11 and 12) show persistence in fund returns: unlike any other asset class, the returns earned historically by venture capital managers are predictive of future performance.

Combine the timing of entry and exit with the variables of different investment strategies, markets selected for investment, and skills of the firm's partners in selecting entrepreneurs and managing investments, and it is not surprising that there are marked differences among funds in how well they do with their investments.

Table 8.1 tracks venture capital funds by their start year ("vintage year") from 1969 to 2005, and breaks out their returns by average, maximum, minimum, and relative ranking of the funds in terms of profitability.

For example, the table shows that of the 63 reporting funds started in 1987, and presumably liquidated by 1997, the top 25% returned an IRR of 17.6%, whereas the lowest performing 25% actually lost 0.3%.

You might think that the poor performers would come much closer to the best funds during periods when everyone is making money. That is not the case. Even the best vintage years show huge differences in results. For example, of the 35 funds started in 1996, the top quartile funds had an IRR of 113.90% whereas the bottom quartile had an IRR of only 1.40%.

This holds true for periods of lower general performance as well. If we look at the 57 funds started in 2001, for example, the top quartile delivered an IRR of 9.60%, while the bottom 25% lost 3.90%.

The spread between best and worst performers during downturns looks smaller, but looks can be deceiving. Funds started in 2001 were not fully liquidated as of 2008. Therefore, the returns shown are actually estimates based on realized profits from investments that have been sold plus projected profits for those still held in the portfolio. We may yet see the gap between top and bottom widen once all the accounting is done.

These data conclusively demonstrate that the profitability of a fund is highly dependent not only on its investment strategy and skill in picking winners, but on the timing of investments and exits relative to public market and general economic conditions.

They also prove what one would intuitively expect: top performers deliver superior return on investment regardless of the environment in which they operate.

THE INVESTMENT HORIZON

Venture capitalists continue to seek profitable investment opportunities, trusting that the market will reward their efforts when the time comes to exit the investments. Their object, as always, is to create and grow companies that exploit innovations with large potential for growth and profit.

It has never been easy to build successful ventures, and it won't get easier. As industries consolidate, barriers to market entry will likely grow higher. Furthermore, big global companies have become much nimbler competitors than they were in the early years of venture capital.

But this does not mean that the doors are closed. As one observer wrote:

> Markets are swampy ... lumps of chaos studded with redoubts and obstacles that disappear and reappear from any direction, studded with over and under ramparts onto which confused invaders stumble and then stagger off from view. And even when you succeed in your objectives and take the field, sometimes the field disappears beneath you, and you find yourself slogging about in pale foam that obscures your vision and leaves you wandering directionless in a vast wilderness.

> And sometimes you get amazingly lucky.[13]

---

[13] M. R. Chapman, *In search of stupidity: Over 20 years of high-tech marketing disasters*, 2nd edn. (New York: Apress, Springer-Verlag, 2006), p. 276.

Technology-driven markets do not remain static for long. They are subject to disruptive surprises from creative entrepreneurs who take advantage of unheralded innovations, changing societal demands, or changes in government regulations to build successful businesses. Opportunities for newcomers will continue to be created by this activity.

Innovation can still trump consolidation

With technologies and global markets changing faster than ever before, industry leaders are more vulnerable to competition.

A study of thirty-five industries by McKinsey & Co. is especially revealing.[14] It shows that companies in the top 20 percent in terms of revenue are having a harder time maintaining that elevated status. In 2002 twice as many of them dropped off the list within 5 years as in the mid-1970s.

During the same period an even more drastic outcome, exit through bankruptcy or merger, actually tripled in frequency. Ten years ago, who would have predicted that General Motors was headed for Chapter 11?

This suggests that even mature markets are always being redefined. Dominant companies are falling from grace at an accelerating rate, pushed aside by competitors with better technologies and shrewd strategies. Market openings are there for smart investors and entrepreneurs who respond to changing conditions or take advantage of errors by market leaders.

Judging from the fact that patent generation in all technologies continues to accelerate (as it has since the days of Commissioner Ellsworth), the world is not running out of inventions or inventors. Corporations, universities, and research institutes continue to generate innovations affecting everything from new software

---

[14] W. J. Huyett and S. P. Viguerie, "Extreme competition," *The McKinsey Quarterly Report*, 1 (2005), 48.

to new energy sources. Some will certainly be well suited for commercialization through venture capital investments.

Identifying potential commercial winners is the big problem, but that has always been true. Recent history shows that the "next big things" are likely to emerge without much warning. Who expected cellular phones to sweep through the world's consumer market as rapidly as they did? How many experts foresaw the success that startups would have in building a host of applications for mobile handsets?

For that matter, remember how a microprocessor from a newcomer called Intel redefined the semiconductor industry, which until then had been dominated by Motorola, Texas Instruments, and a few other big companies? Or how Google came to dominate the Internet industry with a business model based on proprietary search technology?

These are familiar examples of new markets created by innovations that were as unexpected as they were revolutionary. In many cases they transformed industries that had looked like oligopolies into vibrant, competitive marketplaces, spawning new opportunities for startups in the process.

Sometimes this happens indirectly. When consolidation of equipment vendors occurred in some fast-growing markets, it only led to the emergence of startups offering applications software that runs on these platform products. The example I have in mind is the wireless industry, where a few vendors produce most of the handsets in the world. But it is primarily small companies that develop the application software. By adding valuable features to the handsets that prove popular with consumers, they greatly increase the popularity of these devices.[15]

I picked this example because innovations built around wireless technologies is a fertile field for entrepreneurs. They are launching new companies to develop and sell novel software not only to

---

[15] E. J. Savitz, "Who's really behind the iPhone success," *Barron's* (June 15, 2009), 37.

enable new consumer services for the wireless industry, but to help the industry improve the efficiency of its infrastructure.

The challenge of growing software companies into big businesses, covered in detail in Chapter 7, hasn't dampened their enthusiasm. Startups have already released more than 25,000 software applications for the iPod. A flood of business plans emerges whenever a new set of market opportunities appears. In 2008 alone one prominent venture fund received 2,500 proposals for potential iPhone application startups.[16]

VMware breaks the data center entry barrier
Wireless application software is basically a consumer market opportunity. Building a new business in mission-critical software products after the market has consolidated is much more difficult. But companies have done it.

I mentioned earlier (Chapter 6) one such champion: BEA Systems, a Warburg Pincus startup that grew into a billion-dollar company by selling mission-critical enterprise software to the biggest global corporations.

More recently, another successful company, VMware (not a Warburg Pincus investment), has become noteworthy for this achievement. Founded in the late 1990s, VMware proved once again that startups can successfully launch a blockbuster new product category in a very difficult environment – if the timing is right.

VMware found its competitive opportunity in a market not normally open to new competition: mission-critical software for data centers. It succeeded through a combination of innovative technology, good marketing strategy, and being in the right place at the right time. And it kept its competitive edge over time.[17]

---

[16] Comment by Matt Murphy, from the venture capital firm of Kleiner Perkins Caufield and Byers, reported in *PEWeek* (September 29, 2008), 7.

[17] J. Hernick, "VMware still tough to beat, but it's pricey," *Informationweek.com* (April 27, 2009), 52.

To understand how difficult this was, recall the "lessons learned" in Chapter 7, where we discussed the hazards of building a big company to sell mission-critical software to large enterprises. Big customers for mission-critical software have always been more comfortable dealing with major vendors such as Computer Associates, Oracle, and IBM for their software needs.

But VMware found an opening in the market that it could exploit. The company's primary product was (and is) proprietary virtualization software which dramatically increases the utilization rate of computers in data centers.

No new equipment or modifications to existing processors are necessary. VMware's product is a software-only addition that lets data centers handle more processing chores with the same equipment, or the same workload with fewer computers.

Its value proposition is compelling: less equipment needed for a given workload. Fewer computers mean lower capital costs. They also meant lower utility bills, an important factor in a period of rising electricity costs. The software found a ready market.

One huge factor in the company's success was its marketing strategy, which dealt head-on with the disadvantage startups have in selling mission-critical products to large enterprises. VMware markets mostly through a large number of system integrators and resellers of software and hardware, which already have relationships with the biggest companies.

With over 10,000 channel partners marketing its products, VMware achieved revenues exceeding $1.8 billion in 2008. The most impressive proof of its success, however, is the fact that its products are found in 90 percent of the Fortune 1000 list of the top enterprises in the US.

VMware not only launched a new product category, now widely adopted and being emulated by a growing number of big and small competitors. It also proved there are still markets and technologies that are less capital-intensive, where smart newcomers can go straight through to successful commercial rollout.

VMware was acquired by EMC in 2004, but was spun out as a separate business with a public market listing in 2007.

HOT MARKET: ALTERNATIVE ENERGY

Innovations in the hands of talented entrepreneurs can generate valuable businesses even in difficult markets. VMware's virtualization software is a perfect example: a new product, with a new application, targeted at developing a new market niche that turned out bigger than was probably anticipated by the founders.

However, evolutionary innovations are unlikely to spark the hot markets and investment waves with which we opened this book. Such markets become multibillion dollars in size and attract billions of dollars of investment capital. They require transformational developments that promise to disrupt industries and remake whole segments of society.

Innovations of that significance are always in short supply, and it is virtually impossible to predict when one will erupt into prominence the way computers, telecommunications, and the Internet did in the pre-2000 period. From our vantage point, as we move into the second decade of the twenty-first century, the most prominent candidates include technologies for renewable sources of energy; associated electrical storage technologies; and technologies for better managing the efficient use of energy.

As we noted in Chapter 1, VCs have recently been pouring money into alternative energy companies. There is certainly enough public and corporate support for a new energy initiative to inspire optimism. During the US presidential campaign of 2008 both candidates treated the development of alternative energy as a major objective for the future. Even in the face of other urgent financial priorities, the new administration says it is committed to funding an alternative energy generation initiative that will "unleash entrepreneurs and create a tidal wave of innovation." Large industrial firms are placing bets on it. GE makes wind turbines. BP, a major oil company, has adopted "beyond petroleum" as an advertising tagline.

For young companies and the VCs that fund them, however, things are not so simple. Alternative energy startups have capital requirements that are very different from anything most VCs have seen before.

We are talking about companies that will be manufacturing products in large volumes. It doesn't matter whether we are looking at biofuels, solar panels, battery technology for residential and vehicular use, fuel cells, or wind turbines. All innovative energy technologies are tied to the development of efficient, high-output production processes.

That makes companies in these industries capital-intensive by definition. They have large fixed costs related to building and maintaining production plants. Their capital needs may well range in the hundreds of millions of dollars.

These numbers dwarf what is normally required to develop companies in technologies most familiar to the venture capital industry, such as software, where entry costs are low, or electronic systems and semiconductors, where outsourcing of production is the norm. Given the size of most venture capital funds, this is a daunting prospect. Their strategy is to fund innovative startups, but rely on other sources of capital for expansion.

VCs have already invested in hundreds of energy startups, and their ultimate capital needs are well understood. Figure 8.9 shows the number of such companies receiving venture capital funding, the total annual funding since 1994, and the amounts invested per deal.

In 2007, for example, over 250 companies were funded at an average of over $15.5 million per company. Obviously, such sums are inadequate to build production plants.

These are seed investments. Investors are counting on getting the early stages of the technology off the ground. Then they plan to raise much larger rounds of funding from sources such as corporations, public markets, and institutions of all kinds. Debt

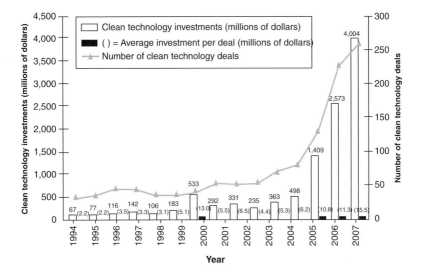

FIGURE 8.9. Clean technology investments between 1994 and 2007.
Source: Based on data from Thomson Reuters.

financing is also sometimes an option because of government-supported guarantees.

In addition to the inherent technology risks, this strategy is clearly subject to fickle financial markets, global competition, and changing government policies. What seems clear is that when it comes to transformational energy technology, the ground for venture investing has shifted. The scale and scope of alternative energy initiatives dictate that venture capital cannot carry the businesses to a successful result on its own.

WHAT ABOUT THE FUTURE?

Several important real-world considerations must guide our thinking about the future of venture capital and VCs.

First, the amount of venture capital that can be *profitably* deployed annually is not infinite. But profitable deployment is not determined by the total amount of investment capital available either. Profits really grow out of the quality of the ideas available. As we look at the future of entrepreneurship and the ability to fund new businesses, I don't think we need to worry about running out of ideas. But in any one year, only a few of them have the potential to spark the creation of really valuable companies differentiated by their products and business strategy.

Indeed, history has shown that too many startups chasing the same market will inevitably produce too many business failures. So the fact that the total amount of funding available has declined since 2000 speaks to a need to reduce the number of startups, while ensuring that follow-on funding is there to finance their growth when public markets may not be open.

Second, the amount of venture capital available is ultimately related to the profitability of venture capital funds. There is no incentive for institutions to put money into the high-risk enterprises that VCs typically finance if they can get the same or better returns from investing in safer market plays, such as bonds or the stock market.

So insofar as venture capital continues to be available, its distribution will be determined by profitability. This reality has profound implications for the size of the venture capital industry as a whole. Profitability is impacted by factors both within and outside the control of venture capitalists. As long as the best-managed VCs and their companies learn to adapt quickly to changing circumstances, and are thus more profitable, they will probably get the lion's share of the money. Less-skilled competitors may find it hard to attract investments. Given the funding situation, they will find it difficult to survive, and so the total number of VCs will decline.[18]

History repeats itself – but in different form

Throughout this book we have focused primarily on the period from 1980 to 2000, which was remarkable for the wealth of opportunities it offered. It is reasonable to ask whether such favorable conditions

---

[18] P-W. Tam, "Venture capitalists head for the door," *The Wall Street Journal* (June 5, 2009), C1.

will ever appear again. I am tempted to answer with another question: why shouldn't they, if we believe that innovations will continue to drive economic growth?

Economic conditions are unpredictable, but there are certain constants we can count on to recur. Business cycles are a way of life. GDP growth reappears periodically. Investment opportunities rarely repeat themselves in the same form, but technological innovations will continue to appear, some of which will have a transformational effect on the world's economy.

Given these basic principles, it seems logical that companies at the forefront of commercializing innovation and creating new markets will continue to build enterprise value. It is also safe to assume that factors that have impacted the success of ventures in the past will continue to be relevant in the future.

In the course of this book we have examined how interesting investment opportunities emerged in several technology-based industries, how we decided which companies to fund, and the outcome of our investment decisions. To illustrate these points we went into some detail on the history of thirteen companies funded by Warburg Pincus, each of which focused on new markets and confronted large potential competitors. We discussed important lessons learned from our successes and disappointments, important to keep in mind when considering future investments.

Being thoughtful about known risk factors, however, cannot shield investments from the influence of the prevailing economic environment. This can act as either friend or foe. For example, one unanticipated risk for investments made in the late 1990s was the decline in capital market interest in young companies after 2000, an effect that ultimately led to the reduced profitability of many venture capital funds of that period. As a result, there has been a sharp decline in the capital available for investments using the "classic" venture capital model of primarily funding startups.

Barring a dramatic increase in general industrial activity, the amount of capital available for early-stage investing will continue to be less than in the frothy 1990s. In reaction to the need for more selectivity, venture capital funds will put a sharper focus on investing in specialized areas where they can add the most value to their portfolio companies.

The disparity in financial returns between the top performing firms and the others will lead to clearer thinking about how VCs add value.

Small funds will focus on startups that aim to demonstrate the commercial viability of new products or technologies. Some of these will continue as free-standing businesses on the basis of new funding. Others will be sold to companies looking for new products and technological capabilities that can be acquired for less than internal development costs. There is nothing new here. This is the conventional approach, and it will persist as long as technologies keep up their rapid evolution.

Large firms with multibillion dollar pools of capital under management have a broader range of investment strategies to choose from. In the case of Warburg Pincus, for example, the quest to create value from companies in all stages of development drives the investment approach. This includes not just initial funding of startups, but what we call "venture growth equity" investing in companies in later stages of development. Such investments could include established businesses that have not lived up to their expectations, or companies that can greatly accelerate their growth with fresh capital.

The reward comes from turning them into industry leaders. In fact, some of these transformational investments might even be in public companies in interesting and growing markets that have been starved of adequate capital to fuel their growth.

The 1980–2000 era was clearly unusual in creating profitable investment opportunities, but history does have a habit of repeating itself in different forms. Public markets periodically bounce back, opening up IPO opportunities for young companies that have demonstrated strong growth, with the business scale to meet industry needs and the potential for market leadership.[19, 20] Venture capital-backed companies will inevitably be among them, having benefited from the unique management skills and technological savvy only a VC can provide.

[19] L. Cowan, "Venture-backed new stocks are back," *The Wall Street Journal* (May 18, 2009), C3.

[20] B. Masters, "Promising IPOs point to flotation thaw," *Financial Times* (May 22, 2009), 15.

Appendix

Table A.1. *Selected industry pioneers (venture or angel capital-backed) that became industry leaders (2008).*

Company	Year founded	IPO year	2004 revenues (millions $)	2008 revenues (millions $)	FY end date	Product
Intel	1968	1971	34,209	37,586	Dec. 27, 2008	Microprocessors
Checkpoint Systems	1969	1977	779	917	Dec. 28, 2008	Security network software
Microsoft[a]	1975	1986	38,474	60,420	Jun. 30, 2008	PC software
Apple Computer	1976	1980	9,763	32,479	Sep. 27, 2008	User-friendly PCs
Oracle[a]	1977	1986	10,557	22,430	May 31, 2008	Relational databases
3-Com	1979	1984	669	1,294	May 30, 2008	Ethernet networks
Seagate	1979	2002	6,129	12,708	Jun. 27, 2008	Disk drives for the masses
Sun Microsystems	1982	1986	11,230	13,880	Jun. 30, 2008	Unix workstations
Electronic Arts	1982	1989	2,957	3,665	Mar. 31, 2008	Electronic computer games
Autodesk	1982	1985	1,234	2,172	Jan. 31, 2008	Design automation software
Cisco Systems	1984	1990	23,579	39,540	Jul. 26, 2008	Computer networks
Xilinx	1984	1990	1,586	1,841	Mar. 29, 2008	Programmable logic chips
Comverse Technology	1984	1986	959	1,867[b]	Jan. 31, 2009	Voice mail systems for telecommunication
Amazon.com	1994	1997	6,921	19,166	Dec. 31, 2008	Internet store
BEA Systems	1995	1997	1,012	1,536	Jan. 31, 2008	Internet transaction software
eBay	1995	1998	3,271	8,541	Dec. 31, 2008	Internet auction service

Table A.1. (cont.)

Company	Year founded	IPO year	2004 revenues (millions $)	2008 revenues (millions $)	FY end date	Product
Covad[c] Communications	1996	1999	429	484[c]	Dec. 31, 2007	Digital Subscriber Loop (DSL) service
Google	1998	2004	3,189	21,796	Dec. 31, 2008	Internet search engine
VMware, Inc.	1998	2007	219	1,881	Dec. 31, 2008	Virtualization product solutions
NeuStar	1999	2005	165	489	Dec. 31, 2008	Telecom intercarrier services
Salesforce.com	1999	2004	96	749	Jan. 31, 2008	On-demand Customer Relationship Management (CRM) services

Notes:

[a] Initial capital from angel investors

[b] Comverse numbers are estimates as they are in the process of restating their historical financial statements

[c] Covad was taken private in April of 2008

Table A.2. *Selected venture capital-backed industry innovators (acquired).*

Company	Year founded	Product	IPO year	Year acquired
Digital Equipment Corporation	1957	Minicomputers	1981	1998
WordStar	1978	Pioneered word processing software for PCs	-	1994
Ortel	1980	Semiconductor lasers for cable systems	1994	2000
Lotus	1982	Spreadsheets for PC	1983	1995
Compaq	1982	First portable PC	1983	2002
Epitaxx	1983	Fiber optic light detectors	-	1999[a]
Level One Communications	1985	Ethernet connectivity chips	1993	1999
Maxis	1987	Games for PCs	1995	1997
Netscape	1994	Internet software	1995	1998
NexGen	1988	Intel compatible microprocessors	1995	1996

Note:

[a]Acquired by JDS Uniphase

Bibliography

Alcaly, R. *The new economy: And what it means for America's future* (New York: Farrar, Straus and Giroux, 2003).

Allen, R. C. *The British industrial revolution in global perspective* (Cambridge: Cambridge University Press, 2009).

Belanger, D. O. *Enabling American innovation: Engineering and the National Science Foundation* (West Lafayette, IN: Purdue University Press, 1998).

Berkery, D. *Raising venture capital for the serious entrepreneur* (New York: McGraw-Hill, 2007).

Bernstein, P. L. *Against the Gods: The remarkable story of risk* (New York: John Wiley & Sons, 1996).

Carlson, C. R. and W. W. Wilmot, *Innovation: The five disciplines for creating what customers want* (New York: Crown Publishing Group, 2006).

Choate, P. *Hot property: The stealing of ideas in an age of globalization* (New York: Alfred A. Knopf, 2005).

Christensen, C. M. *The innovator's dilemma: The revolutionary book that will change the way you do business* (New York: Collins Business Essentials, 2003).

Christensen, C. M. *The innovator's solution: Creating and sustaining successful growth* (Cambridge, MA: Harvard Business School Press, 2003).

Collins, J. *Good to great: Why some companies make the leap ... and others don't* (New York: Harper Business, 2001).

Committee on Science, Engineering, and Public Policy, *Rising above the storm: Energizing and employing America for a brighter economic future* (Washington, DC: The National Academy Press, 2007).

Estrin, J. *Closing the innovation gap* (New York: McGraw-Hill, 2008).

Foster, R. and S. Kaplan, *Creative destruction: Why companies that are built to last underperform the market – and how to successfully transform them* (New York: Currency, 2001).

Friedman, T. L. *The world is flat: A brief history of the twenty-first century* (New York: Farrar, Straus and Giroux, 2005).

Fuerst, O. and U. Geiger, *From concept to Wall Street: A complete guide to entrepreneurship and venture capital* (New York: Financial Times Prentice Hall, 2003).

Gladstone, D. and L. Gladstone, *Venture capital investing: The complete handbook for investing in private businesses for outstanding profits* (New York: Financial Times Prentice Hall, 2003).

Gompers, P. A. and J. Lerner, *The money of invention: How venture capital creates wealth* (Boston, MA: Harvard University Press, 2001).

Graham, B. *The intelligent investor*, revised edition (New York: Collins Business Essentials, 2006).

Gupta, U. (ed.). *Done deals: Venture capitalists tell their stories* (Boston, MA: Harvard Business School Press, 2000).

Hagel, J. H. III and J. S. Brown, *The only sustainable edge: Why business strategy depends on productive friction and dynamic specialization* (Boston, MA: Harvard Business School Press, 2005).

Hansen, A. H. *Business cycles and national income* (New York: W.W. Norton & Company, 1951).

Hargadon, A. *How breakthroughs happen: The surprising truth about how companies innovate* (Boston, MA: Harvard Business School Press, 2003).

Hecht, J. *City of light: The story of fiber optics* (New York: Oxford University Press, 1999).

Josephson, M. *Edison, a biography* (New York: History Book Club, 1959).

Kawasaki, G. *The art of the start: The time-tested, battle-hardened guide for anyone starting anything* (New York: Portfolio, 2004).

Kressel, H. and T. V. Lento, *Competing for the future: How digital innovations are changing the world* (Cambridge: Cambridge University Press, 2007).

Lamoreaux, N. R. and K. L. Sokoloff, with foreword by W. H. Janeway, *Financing innovation in the United States: 1870 to the present* (Cambridge, MA: The MIT Press, 2007).

Lerner, J., F. Hardymon, and A. Leamon, *Venture capital and private equity: A casebook,* third edition (New York: John Wiley & Sons, 2005).

McCraw, T. K. *Prophet of innovation: Joseph Schumpeter and creative destruction* (Boston, MA: Harvard University Press, 2007).

MacVicar, D. and D. Throne, *Managing high-tech start-ups* (Boston, MA: Butterworth-Heinmann, 1992).

Metrick, A. *Venture capital and the finance of innovation* (New York: John Wiley & Sons, 2006).

Mokyr, J. *The lever of riches: Technological creativity and economic progress* (New York: Oxford University Press, 1990).

Naughton, J. *A brief history of the future: The origins of the Internet* (London: Weidenfeld & Nicolson, 1999).

Owen, D. *Copies in seconds: Chester Carlson and the birth of the Xerox machine* (New York: Simon & Schuster, 2004).

Perez, C. *Technological revolutions and financial markets: The dynamics of bubbles and golden ages* (Cheltenham, UK: Edgar Elgar, 2002).

Poser, T. *The impact of corporate venture capital: Potentials of competitive advantages for the investing company* (Wiesbaden, Germany: Deutsche University-Verlag GmbH, 2003).

Prestowitz, C. *Three billion new capitalists: The great shift of wealth and power to the East* (New York: Perseus Books Group, 2005).

Riordan, M. and L. Hoddeson, *Crystal fire: The birth of the information age* (New York: W.W. Norton & Company, 1997).

Schumpeter, J. A. *Capitalism, socialism, and democracy* (New York: Harper Perennial, 1975).

Schumpeter, J. A. *The theory of economic development* (New Brunswick, NJ: Transaction Publishers, 2008).

Sull, D. N. *Made in China: What Western managers can learn from trailblazing Chinese entrepreneurs* (Boston, MA: Harvard Business School Press, 2005).

Thompson, D. G. *Blueprint to a billion: 7 essentials to achieve exponential growth* (New York: John Wiley & Sons, 2006).

Utterback, J. M. *Mastering the dynamics of innovation* (Boston, MA: Harvard Business School Press, 1996).

Walsh, J. *Keynes and the market* (New York: John Wiley & Sons, 2008).

Womack, J. P. and D. T. Jones, *Lean solutions: How companies and customers can create value and wealth together* (New York: Free Press, 2005).

Index